*Fish
of
Greece*

FISH
of
GREECE

by George Sfikas

EFSTATHIADIS GROUP

EFSTATHIADIS GROUP S.A.

14, Valtetsiou Str.
106 80 Athens
Tel: (01) 5254650, 6450113
Fax: (01) 5254657
GREECE

ISBN 960 226 198 6

© **Efstathiadis Group A.E. 1998**

Printed and bound in Greece

Contents

SARDINE: The Sardine is very common in Greek waters. It is found in the open sea, swimming in large shoals near the surface. Sardine are generally eaten fried or salted. Length up to 20 cms.

ΣΑΡΔΕΛΑ: Είδος πολύ κοινό στίς ἑλληνικές θάλασσες. Ζεῖ στό ἀνοιχτό πέλαγος σχηματίζοντας μεγάλα κοπάδια πού κολυμποῦν κοντά στήν ἐπιφάνεια. ῾Η σαρδέλα τρώγεται τηγανητή ἤ παστή. Μῆκος μέχρι 20 ἑκ.

SARDINE: Espèce très commune dans les eaux grecques. Elle vit en pleine mer, en bancs très denses qui nagent près de la surface. La sardine se consomme frite ou salée.

SARDELLE: Die Sardine kommt in griechischen Gewässern sehr häufig vor. Sie schwimmt in großen Schwärmen im offenen Meer nahe der Meeresoberfläche. Die bis zu 20 cm lang werdenden Fische werden in Griechenland in der Regel gebraten oder gesalzen gegessen.

SARDINA: Specie comunissima nelle acque greche. Vive in mare aperto in banchi fittissimi che nuotano sfiorando la superficie. Pesce che viene preparato fritto o salato.

SARDIN: En mycket vanlig slags havsfisk i Grekland. Den lever i öppna havet, där den bildar stora svärmar som simmar nära ytan. Sardinen äts i Grekland vanligtvis stekt eller inlagd i saltlake. Den är upp till 20 cm. lång.

SARDIEN: De sardien is zeer algemeen in de Griekse wateren. Hij leeft in open zee in grote scholen dicht onder de oppervlakte. In het algemeen wordt de sardien gebakken of gezouten gegeten. Lengte tot 20 cm.

イワシ （サルデラ）

ギリシャの海では 大へん一般的な魚で、外海で見られる。　海面に近く大きな群をなして泳ぐ。　一般に油で揚げたり塩漬けにしたりして食べる。　体長20cm以下

1. *Sardina Pilchardus Sarthélla*

SEA TROUT: A fresh-water species, the Sea Troutis to be found in the cold waters of the rushing mountainous rivers of northern Greece, and in the artificial mountain lakes now created by the construction of dams. It differs from the aquarium Sea Trout found in the markets. In Greece, Sea Trout are eaten prepared in a variety of ways. Usual length up to 30 cms.; occasionally found up to 60 cms..

ΠΕΣΤΡΟΦΑ: Εἶδος τῶν γλυκῶν νερῶν. Ζεῖ στά ὀρεινά ποτάμια τῆς Βόρειας ᾿Ελλάδας μέ τά κρύα καί ὁρμητικά νερά καθώς καί στίς λίμνες πού δημιουργήθηκαν ἀνάμεσα στά βουνά ἀπό τά φράγματα. Διαφέρει ἀπό τίς πέστροφες τῶν ἰχθυοτροφίων πού βλέπουμε στήν ἀγορά. Τρώγεται μέ διάφορους τρόπους. Μῆκος συνήθως μέχρι 30 ἑκ. καί σπάνια μέχρι 60 ἑκ.

TRUITE: Cette truite d'eau douce se trouve dans les torrents aux eaux froides et impétueuses, ainsi que dans les lacs artificiels formés par les digues entre les montagnes. Cette truite diffère de la truite d'élevage offerte au marché. Sa chair se consomme cuisinée de plusieurs façons. Sa longueur atteint les 30 cm, ou, exceptionnellement, 60 cm.

FORELLE: Als Süßwasserart kommt die Seeforelle in den kalten Gewässern der rauschenden Bergflüsse von Nordgriechenland und auch in den künstlichen Seen (Stauseen) vor. Sie unterscheidet sich von der gezüchteten Seeforelle, die auf den Fischmärkten angeboten wird. Normalerweise wird sie bis zu 30 cm lang, gelegentlich werden aber auch Fische mit einer Länge von 60 cm gefangen. In Griechenland isst man die Seeforelle in den verschiedendsten Zubereitungen.

TROTA: Questo pesce d'acqua dolce si trova nei torrenti tra le cui acque fredde ed impetuose si destreggia. Lo si trova pure nelle acque dei laghi artificiali, tra i monti, formati da dighe. Questa trota differisce da quella di allevamento offerta in commercio. Viene cucinata in vari modi. La sua lunghezza raggiunge i trenta centimetri e eccezionalmente i sessanta.

LAXÖRING: En fisk som lever i sötvatten. Den lever i bergsfloderna i norra Grekland, i det kalla och forsande vattnet och i sjöarna, som bildas mellan bergen av dammbyggen. Den skiljer sig från forellerna, uppfödda i fiskdammar, som vi köper i affärer. Den lagas till på olika sätt. Den är vanligen upp till 30 cm. lång och någon gång ända upp till 60 cm. lång.

FOREL: Een zoetwater-soort die voorkomt in het koude water van de snelstromende bergrivieren van Noord-Griekenland, als ook in de tegenwoordig door de constructie van dammen gecreeerde kunstmatige bergmeren. Deze soort verschilt van de gekweekte forel die men op de markten ziet. Forel wordt in Griekenland op verschillende manieren toebereid. Gewoonlijk tot 30 cm. lang, soms tot 60 cm lang.

マス　（ペストロファ）

北部キリシャの山の急流の冷たい水や、近年のダム建設によって造られた山間の人工湖に見られる。 魚屋で見られる養殖ものとは、同じマスでも種類が異なる。 キリシャではマスは様々な方法で料理される。 通常は体長30cm以下であるが、時として60cmもの大きさのものもいる。

2.   *Salmo Trutta Péstrofa*

PIKE: A flesh-eating fish, the Pike is found in the lakes of northern Greece, and in the river Evro. It is commonly known as the fresh-water "shark", since it will feed on any kind of fish, frogs and small aquatic birds. The flesh is very tasty, and is usually eaten salted or smoked. The eggs are used to produce a kind of caviar. Weight 10 to 15 kg..

TOYPNA: Ψάρι σακροφάγο πού ζεῖ στίς λίμνες τῆς Βόρειας Ἑλλάδας καί στόν ποταμό "Ἕβρο. Τόν λένε καρχαρία τοῦ γλυκοῦ νεροῦ γιατί τρώει τό κάθε τί ἀπό ψάρια μέχρι βατράχους καί μικρά ὑδρόβια πουλιά. Τό κρέας του εἶναι πολύ νόστιμο. Συχνά τόν κάνουν παστό ἤ καπνιστό. Εἶναι γνωστός καί μέ τά ὀνόματα Λοῦτσος ἤ Γουβλί. Ἀπό τά αὐγά του γίνεται ἕνα εἶδος χαβιάρι. Βάρος μέχρι 10 ἤ καί 15 κιλά.

BROCHET: Poisson carnassier qui vit dans les lacs de la Grèce du Nord et dans le fleuve Evros. On l'appelle requin d'eau douce, car il est si vorace qu'il engloutit, en plus des gardons qui constituent sa nourriture habituelle, des grenouilles, des canetons, des rats. Sa chair est succulente, et souvent on la consomme salée ou fumée. De ses oeufs on fait une espèce de caviar. Poids: jusqu'à 10-15 kg.

HECHT: Diesen fleischfressenden Fisch trifft man in den nordgriechischen Seen und im Evros-Fluß. Er ist allgemein als "Hai des Süßwassers" bekannt, da er alle Arten von Fischen, Frösche und auch kleine Wasservögel frißt. Das Fleisch des Hechtes ist sehr schmackhaft und wird gewöhnlich gesalzen oder geräuchet gegessen. Aus seinen Eiern stellt man eine Art Kaviar her. Er wird 10-15 kg schwer.

LUCCIO: Pesce carnivoro che vive nei laghi della Grecia settentrionale e nelle acque del fiume Evros. Lo si chiama luccio d'acqua dolce, siccome è ghiottissimo di anitroccoli, che costituiscono il suo nutrimento abituale, di rane e di ratti che inghiottisce voracemente. La sua carne è succulente e non di rado viene consumato salato o affumicato. Con le sue uova viene preparato una specie di caviale. Raggiunge il peso di 10-15 kg.

GÄDDA: En köttätande fisk, som man finner i sjöarna i norra Grekland och i floden Evros. Man kallar den för sötvattens haj, därför att den äter allt ifrån fiskar till grodor och små vattendjur. Dess kött är mycket gott. Man lagar ofta till den rökt eller inlagd i saltlake. Av dess ägg gör man en slags kaviar. Den väger mellan 10 och 15 kilo.

SNOEK: Een vleesetende vis, die de voorkomt in de meren van Noord-Griekenland en in de rivier de Evros. Hij wordt in het algemeen de "zoetwaterhaai" genoemd omdat hij van alles eet, vissen, kikkers en kleine soorten watervogels. Zijn vlees is zeer smakelijk, en wordt gewoonlijk gerookt of gezouten gegeten. Van de eieren wordt een soort kaviaar gemaakt. De snoek bereikt een gewicht van 10 tot 15 kg.

カワカマス （ツルナ）

北部ギリシャの湖や、エフロ川でとれる食用魚。各種の魚、蛙、小さな水鳥をエサとするため、淡水の「ワカ」にして一般に知られる。肉は大へん美味さ。ふつうは塩漬けやくんせいにされ、その卵にある種のキャビアを作るのに用いられる。ギリシャでは、ツルナの他に、ルツオスともグブリとも呼ばれる。重さ10～15kg。

3. *Esox Lucius Toúrna*

COMMON OR BRONZE BREAM: The Common or Bronze Bream is a fresh-water species, which lives in the lakes and rivers of Macedonia and Thessaly. It prefers tranquil, warm waters. In Greece it is also known by the names Léstia and Létsika. The flesh is not very tasty and is particularly bony. Length up to 50 cms.

ΧΑΝΙ: Είδος τῶν γλυκῶν νερῶν. Ζεῖ σέ λίμνες καί ποταμούς τῆς Μακεδονίας καί τῆς Θεσσαλίας. Προτιμᾶ τά ἤρεμα καί ζεστά νερά. Εἶναι γνωστό καί μέ τά ὀνόματα Λέστια καί Λέτσικα. Τό κρέας του ὄχι πολύ νόστιμο καί μέ πολλά κόκαλα. Μῆκος ὥς 50 ἑκ.

BREME: Espèce d'eau douce. On la trouve dans les lacs, étangs et cours d'eau de la Macédoine et la Thessalie. Elle préfère les eaux calmes et chaudes. En Grèce, elle est connue également sous les noms Lestia et Letsika. Sa chair n'est pas très bonne, et les épines y abondent. Longueur: jusqu'à 50 cm.

BRASSE: Die gewöhnliche oder Bronzene Brasse ist ein Süßwasserfisch, der in den Seen und Flüssen Mazedoniens und Thessaliens vorkommt. Er bevorzugt ruhige, warme Gewässer. In Griechenland ist dieser bis zu 50 cm lang werdende Fisch auch unter den Namen Lestia und Letsika bekannt. Sein Fleisch ist nicht sehr schmackhaft und besonders grätenreich.

PAGELLO: Pesce d'acqua dolce. Lo si trova nei laghi, negli stagni e corsi d'acqua della Macedonia e della Tessaglia. Preferisce le acque tranquille e calde. In Grecia è pure conosciuto sotto la denominazione di "Lestia" e di "Letsika". La sua carne non è tanto buona e vi abbondano le spine. La sua lunghezza raggiunge i cinquanta centimetri.

BRAXEN: En fisk som lever i sötvatten, i sjöar och floder i Makedonien och Thessalien. Den föredrar lugna och varma vatten. Den kallas även för Léstia och Létsika i Grekland. Dess kött är inte så gott och med många ben. Längd upp till 50 cm.

BRASEM: Een zoetwatervis, voorkomend in de meren en rivieren van Macedonië en Thessalië. Hij geeft de voorkeur aan rustig en warm water. Hij is in Griekenland bekend onder de namen Háni, Léstia en Létsika. Het vlees is niet zo smakelijk en is bijzonder graterig. Lengte tot 50 cm.

オオブナ (ハ=)

マケドニアや テッサリーの湖沼に 棲息する淡水魚で. 静かな暖かい水を好む。
ギリシャでは レスティア 又は レツィカとも呼ばれている。　肉は骨が多くてさほど
おいしくない。　体長50cm以下.

4. *Abramis Brama Hani*

CARP: A fresh-water fish, famous for its longevity. Though Asiatic in origin, the Carp has been reported in Greece since ancient times, living in many different lakes. You may also hear Greeks call this fish Kárpos, Sazáni or Griváthi. Its flesh is somewhat tasteless. Weight up to 8 kg., sometimes more.

ΚΥΠΡΙΝΟΣ: Ψάρι τῶν γλυκῶν νερῶν φημισμένο γιά τή μακροβιότητά του. Ἡ καταγωγή του εἶναι ἀσιατική. Ἔχει εἰσαχθεῖ στήν Ἑλλάδα ἀπό παλιά καί ζεῖ σέ διάφορες λίμνες. Εἶναι γνωστό καί μέ τά ὀνόματα Κάρπος, Σαζάνι καί Γριβάδι. Τό κρέας του μέτριο. Βάρος ὥς 8 κιλά ἤ καμιά φορά καί παραπάνω.

CARPE: Poisson réputé pour sa longévité (30 à 40 ans en liberté et plus de 100 en bassin). Son origine est asiatique. Importée en Grèce depuis l'antiquité, elle vit dans les lacs se nourrissant de petits animaux et, plus rarement, de plantes. En Grèce, ce poisson est connu aussi comme Karpos, Sazani ou Grivathi. Sa chair a un goût un peu fade. Poids: jusqu'à 8 kg. quelque fois plus.

KARPFEN: Ein Süßwasserfisch, der wegen seiner Langlebigkeit bekannt ist. Obwohl von asiatischer Herkunft, ist der Karpfen in den verschiedenen griechischen Seen seit dem Altertum zu finden. Er wirt auch Karpov, Sazani oder Grivatsi genannt. Sein Gewicht kann 8 kg und mehr erreichen, aber sein Fleisch hat wenig Geschmack.

CARPA: Pesce noto per la sua longevità (30-40 anni se in libertà, e più di cent'anni in bacino). La sua origine è asiatica. Importato in Grecia fin dalla più remota antichità, vive nei laghi e si ciba di pesciolini e di animaletti e, raramente, di piante. In Grecia è pure conosciuto sotto la denominazione di "Karpos", "sazani" o di "Grivathi". La sua carne non sa di molto, è piuttosto insipida. Raggiunge il peso di otto Kg. Qualche volta un po' di più.

KARP: En sötvattens fisk, känd för att den lever så länge. Dess härkomst är asiatisk. Den har förts in i Grekland för länge sedan och finns i olika sjöar. Den är även känd med namnen Kârpos, Sazáni och Grivadi. Dess kött är tämligen gott. Den väger i allmänhet upp till 8 kilon eller någon gång även mer.

KARPER: Een zoetwatervis bekend om zijn lange levensduur. Uit Azië afkomstig is hij reeds in de oudheid in Griekenland ingevoerd, waar hij in tal van meren voorkomt. Hij is in Griekenland bekend onder de namen Kyprînos, Kápros, Sazáni en Griváthi. Zijn vlees heeft weinig smaak. Gewicht tot 8 kg., soms nog zwaarder.

コイ (キプリノス)

長命なことで知られる淡水魚。、 元々はアジア産であるが、古代からギリシャの 多くの湖で見られていたようである。 又の名をカルポス、サザニ、又はグ リバティと呼ばれる。 肉はあまりおいしくない。 ふつうは8kg止まりだが 時には それ以上のものもいる。

5. *Cyprinus Carpio Kiprinos*

POOR COD: This member of the cod family is similar to the Atlantic Hake, but much smaller. It lives in the open sea in deep waters, and is found up to a depth of 300 metres. Generally eaten boiled. Length not exceeding 35 cms..

ΜΠΑΚΑΛΙΑΡΑΚΙ: Είδος παρόμοιο μέ τόν μεγάλο Μπακαλιάρο τοῦ ᾿Ατλαντικοῦ ἀλλά πολύ μικρότερο. Ζεῖ στήν ἀνοιχτή θάλασσα σέ νερά βαθιά μέχρι 300 μ. Τρώγεται συνήθως βραστό. Τό μῆκος του δέν ξεπερνᾶ τά 35 ἑκ.

CAPELAN: Ce membre de la famille des merlus est apparenté au merlu de l'Atlantique, mais il est beaucoup plus petit. Il vit en pleine mer, se tenant entre 200 et 300 mètres de profondeur. On le mange surtout bouilli. Sa longueur ne dépasse pas les 35 cm.

KABELJAU: Dieser der Kabeljaufamilie angehörende Fisch ähnelt dem atlantischen Hechtdorsch. Er lebt im offenen Meer und man findet ihn bis zu einer Tiefe von 300 m. In Griechenland wird der 35 cm lange Kabeljau gekocht gegessen.

MERLUZZO CAPELLANO: Questo pesce è apparentato al merluzzo dell'Atlantico, con la differenza ch'è di minori dimensioni. Vive in mare aperto, mantenendosi ad una profondità di 200-300 metri. Lo si mangia soprattutto lesso. La sua lunghezza non supera i 35 centimetri.

LITEN TORSK: Denna medlem i Torskfamiljen liknar tämligen den stora torsken, men är mycket mindre. Den lever mitt ute i havet med djup upp till 300 m. Vanligen äts den kokt. Den blir inte större än 35 cm lång.

KLEINE KABELJAUW: Dit lid van het geslacht der kabeljauwen lijkt op de grote kabeljauw van de Atlantische Oceaan, maar is veel kleiner. Hij leeft in open zee in diep water tot 300 m. Hij wordt als regel gekookt gegeten. Lengte niet meer dan 35 cm.

コダラ (バカリアラキ)

このタラ科の魚は. 大西洋のタラと似ているが. ずっと小さい。 外海の海深300mまでの部分に住む。 一般に煮て食べる。 体長35cm以下.

6.   *Gadus Minutus Bakaliaráki*

HAKE: An open-water species, the Hake is to be found living in deep waters, and is very common around Greece. It has a delicious flavour, and is served either boiled, fried or in soup. Length up to 80 cms..

ΜΠΑΚΑΛΙΑΡΟΣ: Είδος πού ζεῖ στήν ἀνοιχτή θάλασσα καί σέ ἀρκετά μεγάλο βάθος. Εἶναι πολύ κοινό στίς ἑλληνικές θάλασσες κι ἔχει νόστιμο ἄσπρο κρέας. Τρώγεται βραστός, σούπα ἤ τηγανητός. Μῆκος μέχρι 80 ἑκ.

MERLU: Poisson pélagique qui se tient à des profondeurs de 100 à 300 mètres, très commun en Grèce. Sa chair blanche est très savoureuse et se consomme bouillie, frite ou en soupe. Longueur: environ 80 cm.

HECHTDORSCH: Ein Fisch, der in offenen, tiefen Gewässern vorkommt und in Griechenland sehr bekannt ist. Er wird bis zu 80 cm lang und sein köstlich schmeckendes Fleisch wird entweder gekocht, gebraten oder in der Suppe serviert.

MERLUZZO: Pesce pelagico che si mantiene a delle profondità di 100-300 metri. Comunissimo in Grecia. La sua carne candida è saporitissima e si consuma lesso, fritto o in zuppa. La sua lunghezza raggiunge gli 80 centimetri circa.

KUMMEL: En fisksort som lever i öppna havet och på ganska stora djup. Den är mycket vanlig i de grekiska haven och har ett gott, vitt kött. Den äts kokt, i soppa eller stekt. Den blir upp till 80 cm lång.

KABELJAUW: Een vis van de open zee, die in diep water leeft, en die zeer algemeen is in de Griekse zeeën. Het vlees is wit en zeer smakelijk en wordt gekookt, gebakken of in soep gegeten. Lengte tot 80 cm.

タラ　（バカリアロス）

外洋種のタラで、深海に住み、ギリシャ全域でよく見られる。　美味で、煮、たり揚げたり、スープに入れたりする。　体長80cm以下。

7. *Merlucius Merlucius Bakaliáros*

JOHN DORY: The John Dory inhabits areas where the bottom is flat and weedy. It usually keeps close to the sea-bed, and frequently buries itself partly in the sand. The flesh is tasty and is eaten prepared in a variety of ways. Also known in Greece as Sanpiéros. Length up to 60 cms..

ΧΡΙΣΤΟΨΑΡΟ: Ψάρι πού ζεῖ σέ ὑποβρύχια λιβάδια. Γενικά εἶναι εἶδος τοῦ βυθοῦ καί συχνά χώνεται στήν ἄμμο μέ τό πλευρό. Ἔχει κρέας ἀρκετά νόστιμο καί τρώγεται μέ διάφορους τρόπους. Λέγεται καί Σανπιέρος. Μῆκος ὥς 60 ἑκ.

SAINT-PIERRE: Le Saint-Pierre habite des eaux au fond riche en végétation. Généralement il se tient près du fond, et très souvent il s'enfouit partiellement dans le sable. Sa chair est savoureuse et on la prépare de plusieurs façons différentes. En Grèce il est aussi appelé Sanpieros. Sa longueur atteint 60-65 cm.

PETERSFISCH: Der Petersfisch lebt in Gewässern mit flachem, grasbewachsenen Boden. In der Regel hält er sich nahe dem Grund auf und gräbt sich oftmals ein. Das Fleisch dieses Fisches ist schmackhaft und wird auf veschiedene Arten zubereitet. In Griechenland ist er auch unter dem Namen Sanpieros bekannt.

PESCE SAN PIETRO: Pesce che vive nel fondo delle acque ricche di vegetazione. Generalmente sta vicino al fondo e non di rado si addentra parzialmente nella sabbia. La sua carne è eccellente e viene cucinata in vari modi. In Grecia è pure conosciuto sotto la denominazione di "Sanpiétros". Raggiunge la lunghezza di 60-65 centimetri.

SANKT PERS FISK: En fisk som lever på platser där botten är platt och gräsrik. I allmänhet en fisk, som lever på djupet och ofta gräver ner sig på sidan i sanden. Den har ett ganska gott kött och äts på olika vis. Man kallar den även Sanpieros. Upp till 60 cm lång.

ZONNEVIS: Deze vis leeft in water met een vlakke en met zeewier begroeide bodem. Gewoonlijk houdt hij zich vlak bij de bodem op, en graaft hij zich gedeeltelijk in het zand in. Het vlees is smakelijk en wordt op verschillende manieren toebereid. In Griekenland is hij bekend als Christópsaro en als Sanpiéros.

マトウダイ （クリストプサロ）

魚が平らで 海草の多い海に棲息する。 ふつうは 海底近くひそんでいて，体の一部を砂の中に埋めていることが多い。 肉は美味で いろいろな料理法で食べられる。 ギリシャでは サンピエロスという名でも知られる。 体長 60cm以下。

8. *Zeus Faber Christópsaro*

GROUPER: *A deep-water type of grouper which prefers sandy or rocky bottoms. It has a pleasant flavour, and is eaten grilled, fried, etc.. It is also served baked in a special Greek dish called "Plaki". Often sold cut into steaks. Length up to 80 cms..*

ΣΦΥΡΙΔΑ: *Εῖδος πού ζεῖ στά βαθιά νερά μέ βυθό ἀμμουδερό, ἤ πετρώδη. Τό κρέας του πολύ νόστιμο. Τρώγεται ψητό, τηγανητό, πλακί κ.λ.π. Συχνά πουλιέται σέ φέτες. Μῆκος μέχρι 80 ἑκ.*

MEROU BLANC: *Le mérou se trouve dans les eaux profondes à fond de sable ou de pierres. Sa chair est très bonne et on peut la manger grillée, frite ou "Plaki", spécialité grecque très appréciée. Souvent elle est vendue coupée en tranches. Sa longueur atteint 80 cm.*

BARSCH: *Ein Tiefseefisch, der bis zu 80 cm lang werden kann und sandige oder felsige Meeresgründe bevorzugt. Sein Fleisch hat einen angenehmen Geschmack und wird gebraten oder gegrillt gegessen. Gebacken ist er als griechische Spezialität unter dem Namen "Plaki" bekannt.*

CERNIA BIANCA: *Pesce che si incontra nelle acque profonde a fondo sabbioso o sassoso. La sua carne è ottima e viene preparata ai ferri, fritta o in umido "Plaki" (specialità greca molto apprezzata dai buongustai). Viene venduto anche affettato. La sua lunghezza raggiunge gli 80 centimetri.*

HAVSABORRE: *En fisk som lever på mycket stora djup med sand- eller stenbotten. Den har mycket gott kött. Man äter den grillad, stekt, eller lagad i ugn "Plaki", en grekisk mycket omtyckt rätt etc. Den säljs ofta skuren i skivor. Den blir upp till 80 cm lång.*

REUZENBAARS: *Een diepzee-vis die voorkeur heeft voor een zand-of steenachtige bodem. Het vlees is aangenaam van smaak en wordt gebakken of geroosterd gegeten, vaak ook op een speciale Griekse manier, "plaki" genoemd. Deze vis wordt vaak in moten gesneden verkocht. Lengte tot 80 cm.*

ハタ （スフィリダ）

ハタ科の深海型で、砂底や岩底を好む。 なかなかおいしい魚で、焼いた
り揚げたりして食べる他、オーブンで焼いて「プラキ」と呼ばれるギリシャ料理
にも使われる。 たいていは切身で売られている。 体長80cm以下。

9. *Epinephelus Aeneus Sfirítha*

BROWN COMBER: A small member of the sea-perch family, found living in shallow waters near the seashore. Prefers a seaweedy, rocky bottom. The flesh is delicious, and is served fried, or boiled together with other fish. The Greeks also call this fish Pérthika or Pérki. Length up to 12 cms..

ΠΕΡΚΑ: Μικρό ψάρι πού ζεῖ στά ρηχά νερά, κοντά στίς ἀκτές. Προτιμᾶ τά βραχώδη μέρη μέ ἄφθονα φύκια. Τό κρέας του εἶναι ἀρκετά νόστιμο. Τρώγεται τηγανητό ἤ βραστό μαζί μέ ἄλλα ψάρια. Λέγεται καί Πέρδικα ἤ Περκί. Μῆκος μέχρι 12 ἐκ.

TAMBOUR: Petit poisson de la famille des perches de mer, le serran préfère les eaux peu profondes, à fond rocailleux avec beaucoup d'algues. Il se tient près des côtes. Assez savoureux, on le mange frit ou bouilli, avec d'autres poissons. Les Grecs l'appellent aussi Perdika ou Perki. Sa longueur atteint 12 cm.

BARSCH: Dieser Fisch gehört zur Seebarschfamilie und lebt in flachen Gewässern nahe der Meeresküste. Sein bevorzugter Aufenthaltsort ist felsiger, mit Seegras bewachsener Meeresboden. In Griechenland wird er auch Perthika oder Perki genannt und sein köstliches Fleisch wird gebraten. Länge: bis zu 12 cm.

PERCA: Pesciolino della famiglia Serranidi dei mari. Preferisce le acque poco fonde, a fondo roccioso coperto d'alghe. Nuoto vicino alle coste. Assai saporito, viene preparato fritto o lesso insieme ad altri pesci. In Grecia è pure ˜onosciuto sotto la denominazione di "Perdika" o "Perki". La ua lunghezza raggiunge i dodici centimetri.

ỈAVSABORRE: En liten fisk som lever i grunt vatten, nära usterna. Den föredrar klippiga platser med fullt med tång. 'ess kött är ganska gott. Man äter det stekt eller kokt med andra fiskar. Den kallas även Pérdika eller Perkï. Dess längd mäter upp till 12 cm.

SERRANUS HEPATUS: Dit kleine lid van de zeebaarzenfamilie leeft in ondiep water langs de kust, bij voorkeur op een met zeewier begroeide rotsige bodem. Het vlees smaakt kostelijk en wordt gebakken of gekookt met andere vis gegeten. De Grieken noemen hem Pérka, Pérthika of Perkï Lengte tot 12 cm.

スズキ （ペルカ）

スズキ科の小魚で、海岸に近い浅瀬に棲息し、海藻の茂る岩底を好む。
肉は美味で、他の魚と一緒に油で揚げたり煮たりして食べる。　ギリシャ人は
ペルティカ、又はペルキとも呼ぶ。　体長12cm以下。

10. **Serranus Hepatus Pérka**

UNSTRIPED RED MULLET: This species is somewhat similar to the Red Mullet, but can be distinguished from it by its blunter nose or profile. It prefers muddy bottoms, and is found up to a depth of 300 metres. The flesh is very tasty, but has a slightly muddy odour. Generally eaten fried. Length 20 to 40 cms..

ΚΟΥΤΣΟΜΟΥΡΑ: Είδος συγγενικό μέ τό μπαρμπούνι πού ξεχωρίζει ὅμως ἀμέσως ἀπ' αὐτό, ἀπό τό κοφτό προφίλ τοῦ κεφαλιοῦ του. Ζεῖ σέ λασπώδεις βυθούς σέ βάθος ὥς 300μ. ῎Εχει πολύ νόστιμο κρέας ἀλλά μέ ἐλαφριά μυρωδιά βούρκου. Γίνεται συνήθως τηγανητό. Μῆκος 20 ὥς 40 ἐκ.

ROUGET-BARBET: Cette espèce ressemble au Rouget de roche (ou Surmulet), dont il se distingue grâce à son profil moins accusé. Il préfère les fonds boueux et on peut le trouver à des profondeurs qui atteignent 300 mètres. Sa chair est bonne, mais avec une certaine odeur de boue. Généralement, ce rouget est servi frit. Longueur: entre 20 et 40 cm.

MEERBARBE: Diese Fischart ähnelt in etwa der Meerbarbe, unterscheidet sich jedoch von dieser durch ihr stumpfes Profil. Sie bevorzugt schlammige Böden und wird bis zu einer Tiefe bis zu 300 m gefunden. Der Fisch hat einen leicht schlammigen Geschmack und wird gebraten gegessen.

TRIGLIA DI FANGO: Questa specie somiglia alla triglia di roccia, tuttavia si distingue dal profilo meno accentuato. Preferisce le acque melmose ove lo si può trovare fino a delle profondità che raggiungono i 300 Metri. La sua carne è succulenta, tuttavia sa di melma. Generalmente fiene servito fritto. Raggiunge la lunghezza di 20-40 centimetri.

RÖDBARB UTAN RÄNDER: Ett fiskslag, släkt med rödbarben, men skiljs dock genast ifrån den genom sin skarpa huvudprofil. Den lever på lerbotten på upp till 300 m djup. Den har mycket gott kött, men med en lätt gyttjesmak. Man äter den vanligen stekt. Den blir mellan 20 och 40 cm lång.

ROODGEBAARDE BARBEEL: Deze vis lijkt enigzins op de rode zeebarbeel, maar heeft een meer stompe snuit. Hij houdt van een modderige bodem en hij komt voor tot een diepte van 300 meter. Het vlees is zeer smakkelijk, hoewel het een ietwat modderige geur heeft. Meestal wordt hij gebakken gegeten. Lengte van 20 tot 40 cm.

シマナシアカホラ （クツォムラ）

アカホラによく似ているが、触かく、特にすしい口先の部分が異なっている。
泥底を好み、海深300mの所で見られる。肉はたへん美味であるが、かすか
に泥臭い。ふつうは油で揚げて食べる。体長20〜40cm。

11. *Mullus Barbatus Koutsomoúra*

RED MULLET: The Red Mullet dwells near seaweedy and rocky bottoms in depths of 50 to 100 metres. Its colour varies according to its environment, but it is generally a reddish-brown. It is a delicious fish, either served fried or grilled. Length 25 to 40 cms..

ΜΠΑΡΜΠΟΥΝΙ: Είδος πού ζεῖ σέ βυθούς βραχώδεις ἤ σκεπασμένους μέ φύκια, σέ βάθος 50 ὥς 100 μ. Τό χρῶμα του ποικίλλει ἀνάλογα μέ τό περιβάλλον. Πάντως γενικά εἶναι κοκκινωπό. Έχει πολύ νόστιμο κρέας καί τρώγεται τηγανητό ἤ ψητό. Μῆκος 25 ὥς 40 ἐκ.

ROUGET DE ROCHE OU SURMULET: Ce Rouget habite près de fonds riches en pierres et algues, se tenant à 50-100 mètres de profondeur. Sa couleur varie suivant l'ambiance, mais, généralement, elle est d'une rouge virant au brun. C'est un poisson délicieux, soit frit soit grillé. Sa longueur est d'environ 25-40 cm.

MEERBARBE: Diese Art wird in einer Tiefe von 50-100 m in der Nähe von Seegras und felsigem Meeresboden gefunden. Ihre Farbe ist gewöhnlich rötlich-braun, ändert sich aber mit der Umgebung. Es handelt sich um einen schmackhaften 25-40 cm langen Fisch, der entweder gebraten oder gegrillt zubereitet wird.

TRIGLIA DI SCOGLIO: Questa triglia vive vicino a fondi sassosi e ricchi di alghe, mantenendosi ad una profondità di 50-100 metri. Il suo colore varia a seconda dell'ambiente, ma generalmente è rosso tendente al marrone. La sua carne è succulentissima sia se mangiato fritto o ai ferri. Raggiunge la lunghezza di 25-40 centimetri.

RÖDBARB: En sort som lever vid klippbotten eller botten täckt med tång på djup mellan 50 och 100 m. Dess färg skiftar efter omgivningen. Men i allmänhet är den dock rödaktig. Den är mycket god och äts stekt eller grillad. Längd mellan 15 och 40 cm.

ZEEBARBEEL: Deze soort leeft op rotsige en met zeegras begroeide bodems op diepten van 50 tot 100 m. De kleur verandert al naar gelang zijn omgeving, maar is in het algemeen roodachtig bruin. Het is een heerlijk smakende vis die gebakken of gegrild wordt gegeten. Lengte van 15 tot 40 cm.

アカボラ （バルブ二 ）

海深50～100mの海草の茂った岩店の近くに棲息する。　色はまわりの状況 に応じて変化するが、ふつうは赤茶色をしている。　おいしい魚で、油で揚けた り、焼いたりして食べる。　体長25～40cm.

12. *Mullus Surmuletus Barboúni*

HORSE MACKEREL: An open-water species, found swimming in large shoals near the surface. The flesh is not strongly flavoured, and is usually eaten fried. Small Horse Mackerel are known as Sampánious, and are salted. Length up to 40 cms..

ΣΑΒΡΙΔΙ: Ψάρι τοῦ ἀνοιχτοῦ πελάγους. Ζεῖ σέ μεγάλα κοπάδια μετακινούμενο κοντά στήν ἐπιφάνεια. Τό κρέας του μέτριο. Τρώγεται συνήθως τηγανητό. Τά πολύ μικρά τά λένε Σαμπανούς καί τά παστώνουν. Μῆκος μέχρι 40 ἑκ.

SAUREL: Espèce pélagique; vit en bancs qui nagent près de la surface. Sa chair, médiocre, se consomme frite. Les individus très petits, appelés Champanious, sont salés. Longueur: jusqu'à 40 cm.

MAKRELE: Eine im offenen Meer vorkommende Abart, die in großen Schwärmen nahe der Oberfläche dahinzieht. Ihr Fleisch ist nicht sehr schmackhaft und wird gewöhnlich gebraten serviert. Kleine Pferdemakrelen sind in Griechenland auch als Sampanioys bekannt; sie werden bis zu 40 cm lang und gesalzen gegessen.

SURO: Specie pelagica. Vive in banchi che sfiorano la superficie delle acque. La sua carne non tanto pregiata viene preparata fritta. I meno grandi vengono salati. Raggiunge la lunghezza di 40 centimetri.

HÄST-MAKRILL: En öppethavsfisk, som lever i stora svärmar, som simmar nära ytan. Dess kött är ganska gott och den äts vanligen stekt. De mycket små fiskarna kallas Sampanous och de lägger man in i saltlake. Längd upp till 40 cm.

HORSMAKREEL: Een vis van de open zee, die in grote scholen dicht onder de oppervlakte zwemt. Het vlees is matig van smaak en wordt gewoonlijk gebakken gegeten. Kleine exemplaren worden in Griekenland Sambanós genoemd, en worden meestal gezouten. Lengte tot 40 cm.

マアジ (サアジ)

外洋種で、海面近く大群をなして遊泳する。 肉はさほど風味がなく、ふつう油で揚げて食べる。 小マアジはソムパニウスと呼ばれ塩漬けにされる。 体長40cm以下.

13. *Trachurus Trachurus Savríthi*

GILD - HEAD BREAM: This species, a member of the sea-bream family, is found living in small shoals in shallow waters. The older members of the species prefer to live alone, inhabiting rocky bottoms or places with seaweed. Many are also to be found in salt-water lakes, as, for example, at Mesolongi. These are much tastier, and are considered a great delicacy. They are usually grilled or fried. This fish is also identified by the Greek names Chrisófa, Kampanári and Kótsa. Length 30 to 60 cms.

ΤΣΙΠΟΥΡΑ: Εἶδος πού ζεῖ σέ μικρά κοπάδια στά ρηχά νερά. Τά πιό ἡλικιωμένα προτιμοῦν τή μοναχική ζωή σέ βραχώδεις βυθούς ἤ σέ μέρη μέ φύκια. Πολλές τσιποῦρες ὑπάρχουν στίς λιμνοθάλασσες ὅπως στό Μεσολόγγι. Αὐτές γίνονται πιό νόστιμες καί εἶναι περιζήτητος μεζές. Τρώγονται ψητές ἤ τηγανητές. Ἄλλα ὀνόματα τῆς τσιπούρας: Χρυσόφα, Καμπανάρι, Κότσα. Μῆκος 30 ὣς 60 ἑκ.

DAURADE: Le spare vit en bancs dans les eaux peu profondes. Les individus plus âgés préfèrent la vie solitaire près des fonds pierreux ou entre les algues. On le trouve aussi dans les lagunes, comme par exemple celle de Messolonghi. La chair particulièrement savoureuse des spares de lagune est une délicatesse très recherchée. Le spare se consomme frit et grillé. En Grèce, on le nomme aussi Chrysofa, Cambanari, Kotsa. Sa longueur est de 30 à 60 cm.

MEERBRASSE: Diese Art der Meerbrassen findet man in kleinen Schwärmen an seichten Stellen des Meeres. Ältere Fische leben allein, und zwar auf felsigem oder mit Seegras bewachsenem Meeresgrund. Auch in Salzwasserseen, wie z.B. in Messolonghi kommen diese Fische vor. Sie sind bedeutend schmackhafter und werden als Delikatesse angesehen. In Griechenland heißen sie Chrisofa, Kampanari oder Kotsa. Sie werden 30-60 cm lang und gegrillt oder gebraten gegessen.

ORATA: Pesce che vive in banchi in acque poco profonde. Ad una certa età preferisce la solitudine cercando rifugio nei fondi sassosi e tra le alghe. Lo si trova pure nelle lagune, come in quelle della città di Missolonghi. L'orata delle lagune è particolarmente saporita e apprezzata dai buongustai. Viene servita fritta o preparata ai ferri. In Grecia questo succulento pesce è pure conosciuto sotto le denominazioni di "Chrysofa", "Cambanari", "Kotsa". La sua lunghezza varia dai 30 ai 60 centimetri.

GOUDBRASEM: Deze soort leeft in kleine scholen in ondiep water. Oudere exemplaren leven solitair op rotsige bodems of op met zeewier begroeide plaatsen. Deze vissen komen ook voor in zee-lagunes, zoals in die bij Mesolóngi. Deze smaken veel beter en worden als een delicatesse beschouwd. Zij worden gebakken of gegrild gegeten. De Grieken noemen deze vis Tsipoúra, en ook Chrysófa, Kambanári en Kótsa. Lengte van 30 tot 60 cm.

タイ科 (ツィプラ)

タイ科に属するこの魚は、浅瀬に群をなして棲息するが、成長すると群から離れるのを好み、海藻の茂る岩店に単独で住む。 メソロンギのような塩水湖にも多数見られ、こちらの方が海のものよりも美味、珍味と言える。 ふつうは焼いたり油で揚げたりして食べる。 クリソファ、カムパナリ、コッツァなどの別名がある。 体長 30〜60 cm.

14. *Sparus Auratus Tsipoúra*

PANDORA: A type of sea-bream which lives not far from the seashore, in relatively shallow waters. Also known in Greece by the names Lythrinári and Mertzáni. The flesh is white, deliciously flavoured and can be served grilled or fried. Length up to 50 cms..

ΛΥΘΡΙΝΙ: Ψάρι πού ζεῖ ὄχι πολύ μακρυά ἀπό τίς ἀκτές καί σέ νερά σχετικά ρηχά. Λέγεται καί λυθρινάρι ἤ Μερτζάνι. Τό κρέας του, ἄσπρο καί πολύ νόστιμο, τρώγεται ψητό ἤ τηγανητό. Μῆκος μέχρι 50 ἑκ.

PAGEOT ROUGE: Ce poisson ne s'éloigne jamais trop des côtes et vit en des eaux relativement peu profondes. On l'appelle également Lythrinari et Mertzani. Sa chair, blanche et très savoureuse, se consomme grillée ou frite. Sa longueur atteint 50 cm.

ROTE MEERBARBE: Diese Art lebt nicht weit von der Küste in relativ seichtem Wasser. In Griechenland ist sie auch unter den Namen Lythrinari und Mertzani bekannt. Das Fleisch ist weiß, von köstlichem Geschmack und wird gegrillt oder gebraten gegessen. Der Fisch wird bis zu 50 cm lang.

FRAGOLINO: Questo pesce mai si allontana dalle coste e vive in acque relativamente poco profonde. Viene pure chiamato "Lythrinari" e "Martzáni". La sua carne bianca e saporosa è succulenta se fritto preparata ai ferri. La sua lunghezza raggiunge i 50 centimetri.

KNORRHANE: En fisk som lever inte så långt ifrån kusten och i ganska grunda vatten. Den kallas även Lithrinári eller Mertzáni. Dess kött, som är vitt och mycket gott, äts grillat eller stekt. Den är upp till 50 cm lång.

ZEEBRASEM: Deze soort leeft niet ver van de kust in betrekkelijk ondiep water. De Grieken noemen hem Lithrini en ook wel Lythrinári of Mertzáni. Het vlees is blank en uitstekend van smaak, en wordt gegrilld of gebakken gegeten. Lengte tot 50 cm.

タイ科 (リスリニ)

海岸に近い比較的の浅瀬に住むタイ科の魚。 ギリシャでは リスリサリ メルツァニなどの名でも知られる 肉は白身で 風味があり 油で揚げた り焼いたりして食べる。 体長ら ㍍以下。

15.   *Pagellus Erythrinus Lithrini*

SEA-BREAM FAMILY: This sea-bream chooses when young to frequent rocky places near the seashore. When more mature, however, it prefers deep waters and sandy bottoms. The flesh is not strongly flavoured, and is eaten grilled or fried. Length up to 45 cms...

ΣΚΑΘΑΡΙ: Είδος πού σέ νεαρή ἡλικία ζεῖ στούς βράχους κοντά στίς ἀκτές. Ἀντίθετα ὅταν μεγαλώσει προτιμᾶ τά βαθιά νερά καί τούς ἀμμώδεις βυθούς. Τό κρέας του εἶναι μέτριο σέ γεύση. Γίνεται ψητό ἤ τηγανητό. Μῆκος μέχρι 45 ἐκ.

BREME COMMUNE: Ce poisson, tant qu'il est jeune préfère les endroits rocheux près des côtes. Dewenu adulte, au contraire, il préfère les eaux profondes et les fonds de sable. sa chair a un goût médiocre; elle est consommée frite ou grillée. Longueur: jusqu'à 45 cm.

BRASSE: Die Jungfische dieser Meerbrassenart wählen gewöhnlich felsige Stellen nahe der Küste. Wenn sie älter werden bevorzugen sie tieferes Wasser und sandige Meeresgründe. Das Fleisch der bis zu 45 cm lang werden Fische schmeckt leicht fade und wird gegrillt oder gebraten.

PAGELLO COMUNE: Questo pesce finchè giovane preferisce i luoghi rocciosi vicino alle coste. Fattosi adulto cerca le acque profonde i fondi sabbiosi. La sua carne non sa di molto e viene preparata o fritta o ai ferri. La sua lunghezza raggiunge i 45 centimetri.

BRAXEN-FAMILJ: En sort av Havsbraxen-familjen som medan den är ung, lever vid klipporna nära kusterna. Däremot, då den blir äldre, föredrar den de djupa vattnen och sandig botten. Dess kött smakar medelgott. Den äts grillad eller stekt, och blir upp till 45 cm. lång.

ZEEKARPER: Deze soort leeft op jeugdige leeftijd als regel bij rotsen dicht bij de kust. Op oudere leeftijd geeft hij voorkeur aan wateren met een zanderige bodem. Het vlees is matig van smaak en wordt gebakken of gegrilld gegeten. Lengte tot 45 cm.

タイ科 （スカザリ）

このタイ科の魚は、幼魚期には、海岸に近い岩の多い場所を好むが、成魚になると深い砂底を好むようになる。 肉はさほど強い風味はなく、焼いたり油で揚げたりして食べる。 体長45cm以下。

16.   Cantharus Cantharus Skathári

COUCH'S SEA-BREAM: This member of the sea-bream family is abundant in Greek waters. It is found in clear unpolluted waters near sheer rocky coasts or open sea reefs. Prefers a depth somewhere between 20 and 200 metres. It is a well-fleshed, tasty fish and can be eaten boiled, grilled etc.. It is also served baked in a special Greek dish called "Plaki". Length 30 to 70 cms.

ΦΑΓΚΡΙ: Ψάρι ἄφθονο στίς ἑλληνικές θάλασσες. Ζεῖ σέ καθαρά νερά, σέ ἀπόμερες βραχώδεις ἀκτές ἤ πελαγίσιες ξέρες. Τό βάθος πού προτιμᾶ κυμαίνεται ἀπό 20 ὡς 200 μ. Ἔχει νόστιμο καί παχύ κρέας. Γίνεται βραστό, πλακί, στό φοῦρνο κ.λ.π. Μῆκος 30 ἕως 70 ἑκ.

PAGRE COMMUN: Poisson que l'on recontre très souvent dans les eaux grecques. Il habite les eaux claires et non-polluées près des côtes abruptes et roxheuses ou près en pleine mer. Il préfère les profondeurs entre 20 et 200 mètres. C'est un poisson gras et savoureux qui peut être consommé bouilli, grillé etc. En Grèce on le prépare aussi en "Plaki", spécialité très appreciée. Longeur: 30 à 70 cm.

ZAHNBARASSE: Dieses mitglied der Meerbrassenfamilie findet man in griechischen Gewässern im Urfluß. Der Fisch lebt im klaren Wasser in der Nähe felsiger Küsten oder Seeriffs und bevorzugt Tiefen von 20-2000 m. Die Zahnbrasse ist fleischig und schmackhaft und wird gekockt oder gegrillt. Gebacken ist sie als "Plaki" in Griechenland bekannt. Ihre Länge beträgt 30-70 cm.

PAGRO: Pesce di casa nell acque greche. Preferisce le acque limpide vicino alle coste rocciose e a precipizio. Lo si incontra tra i 20 e i 200 metri di profondità. La sua carne è grasse e saporosa. Lo si può mangiare lesso o cucinato ai ferri. In Grecia vienen pure preparato in umido "Plaki", specialità molto apprezzata. la sua lunghezza varia tra i 30 e i 70 centimetri.

HAVSBRAXEN: En vanlig fiskort i grekiska vatten. Man finner den i rent, oförorenat vatten, vid branta klippiga stränder eller öppet havsrev. Djupet, som den föredrar att leva på, växlar mellan 20 till 200 m. Den har ett gott och fett kött. Man tillagar den kokt, "plaki", i ugnen etc. Den blir mellan 30 och 70 cm lång.

PAGRUS: Deze soort komt in de Griekse zeën in grote aantallen voor. Hij leeft in helder, niet vervuild water in de buurt van steile rotskusten of bij riffen in open ee op diepten van 20 tot 200 m. Het is een vlezige en smakelijke vis, die gekookt, geroosterd of gestoofd kan gegeten worden. Ook wordt hij in Griekenland op een speciale Griekse manier, die "plaki", wordt genoemd, gebakken. Lengte 30 tot 70 cm.

タイ科（ファグリ）

このタイ科の魚はギリシャの海で豊富に見られる。 けわしい岩海に近いきれいで汚染されていない水や、外海に面した岩礁で見られる。 海深20〜200mの所を好む。 肉の厚いおいしい魚で、煮たり焼いたりして食べる。 オーブンで焼かれて「プラキ」というギリシャ料理にも登場する。 体長30〜70cm。

17. *Pagrus Pagrus Fagrí*

DENTEX: This sea-bream is found far out from the shore, living near open-water reefs in a depth of 20 to 40 metres. It is exceptionally good to eat, and is generally found boiled or grilled in steaks. Length 50 cms. to 1 metre.

ΣΥΝΑΓΡΙΔΑ: Είδος πού ζεϊ μακριά άπό τίς άκτές σέ πελαγίσιες ξέρες (σκόπελους, ύφάλους) καί σέ βάθος 20 ώς 40 μ. Έχει κρέας έξαιρετικά νόστιμο καί γίνεται συνήθως βραστή ή ωητή σέ φέτες. Μήκος 50 έκ. ώς 1 μ.

DENTE: On rencontre le denté assez loin des côtes, près des récifs de pleine mer, se tenant à une profondeur de 20 à 40 mètres. Sa chair est exceptionnellement savoureuse et, généralement, on le consomme bouilli ou grillé en tranches. Longueur: entre 50 cm et 1 mètre.

ZAHNFISCH: Diese Seebrassenart lebt weit von der Küste entfernt im offenen Meer in der Nähe von Riffs und in einer Tiefe von 20-40 m. Sie wird 50-100 cm lang und wird gekocht oder als Steaks gegrillt.

DENTICE: Pesce che si incontra lontano dalle coste, in mare aperto. Si mantiere a profondità varianti tra i 20 ed i 40 metri. Pesce eccezionalmente saporito viene consumato o lesso o affettato e arrostito ai ferri. La sua lunghezza varia tra i 50 ed o 100 centimetri.

HAVSRUDA: En fisk som lever långt ifrån kusterna vid havsrev och skär, och på 20 till 40 m djup. Den har ett utsökt gott kött, och man lagar vanligtvis till den kokt eller grillad i skivor. Längden är från 50 cm upp till 1 m.

TANDBRASEM: Deze soort leeft ver buiten de kust bij riffen in open zee op een diepte van 20 tot 40 m. Dee vis is buitengewoon smakelijk en wordt meestal in moten gekookt of geroosterd. Lengte van 50 tot 100 cm.

タイ科 (シナグリダ)

このタイ科の魚は 岸から遠く離れた外海の岩礁付近、海深 20~40mの
所に棲息する。 絶妙なおいしさで、ふつうは切り身を煮たり焼いたりして
食べる。 体長 50cm ~ 1 m。

18. *Dentex Dentex Sinagrítha*

SADDLED BREAM: This small member of the sea-bream family lives in shoals, and frequents clear, unpolluted, deep waters wherever the bottom is sandy or covered with seaweed. The flesh is very appetizing, and can be eaten fried or grilled. Length 20 to 30 cms.

ΜΕΛΑΝΟΥΡΙ: Μικρό ψάρι πού ζεῖ σχηματίζοντας κοπάδια σέ μέρη μέ νερά καθαρά καί βαθιά, ἀλλά μέ βυθό ἀμμουδερό ἤ σκεπασμένο μέ φύκια (φυκιάδεσ). Τό κρέας του πολύ νόστιμο. Γίνεται τηγανητό ἤ ψητό. Μῆκος 20 ὥς 30 ἑκ.

OBLADE: Ce petit poisson vit en bancs en eau claire et profonde à fond de sable ou d'algues. Sa chair savoureuse peut être servie frite ou grillée. Longueur: 20 à 30 cm.

OBLADA MELANURA: Das 20-30 cm lange Mitglied der Meerbrassenfamilie bevorzugt klares, tiefes wasser mit sandigem oder mit Seegras bewachsenem Grund. Das Fleisch ist wohlschmeckend und wird gebraten oder gegrillt.

OCCHIATA: Questo pesce preferisce le acque poco profonde a fondo roccioso o sassoro. Pesce comunissimo nelle acque greche. La sua carne è particolarmente succulenta in autunno. Lo si può cucinare in diversi modi: fritto, lesso ecc. E' conosciuto in Grecia anche sotto la denominazione di "Cacaréllos" e di "Caraghiózis". La sua lunghezza varia tra i 15 ed i 25 centimetri.

HAVSBRAXEN-FAMILJEN: En liten fisk som lever i svärmar vijd platser med rent och djupt vatten, och med sandbotten eller täckt av tång. Dess kött smakar mycket gott. Den tillagas stekt eller grillad. Längden är mellan 20 och 30 cm.

OBLADA MELANURA: Een kleine soort zee brasem die in scholen leeft in helder, diep water met een zanderige of met zeewier begroeide bodem. Het vlees smaakt zeer goed en wordt gebakken of gegrilld gegeten. Lengte 20 tot 30 cm.

タイ科 (メラヌリ)

タイ科のこの小さな魚は 浅瀬に住むが 底が砂だったり海草が茂っていれば 水のきれいな深海でもよく見られる。 肉は大へんに美味で、油で揚げたり、 焼いたりして食べる。 体長 20〜30cm.

19. *Oblada Melanura Melanoúr*

ANNULAR BREAM: A type of sea-bream which prefers shallow waters and a rocky or stony bottom. Very common around Greece. The flesh is particularly tasty in autumn, and is served grilled, fried or boiled. Also known in Greece by the names Kakaréllos, Karangiózis or Krachantzis. Lengthn15 to 25 cms.

ΣΠΑΡΟΣ: Ψάρι πού προτιμᾶ τά ρηχά νερά καί τούς βραχώδεις ἤ πετρώδεις βυθούς. Πολύ κοινό στίς ἑλληνικές θάλασσεσ. Τό κρέας του ἰδιαίτερα νόστιμο τό φθινόπωρο. Γίνεται ψητός, τηγανητός, βραστός. Λέγεται καί Κακαρέλλος, Καραγκιόζης ἤ Κραχαντζῆς. Μῆκος 15 ὥς 25 ἑκ.

SPARAILLON A QUEUE NOIRE: Ce poisson préfére les eaux peu profondes à fond rocailleux ou pierreux. Il est très commun dans les eaux grecques. Sa chair est particulière-ment savoureuse en automne. Elle peut étre grilée, frite ou bouilie. Ce poisson, également appelé Cacarellos, Caraghio-zis ou Crahatzis, mesure 15 à 25 cm.

MEERBRASSE: Eine Meerbrassenart, die seichte Gewässer und felsigen Grund bevorzugt. Dieser Fisch ist in Griechen-land sehr häfig. Er wird 15-25 cm lang und sin Fleischt ist besonders im Herbst sehr schmackhalt. Man iβt ihn gebraten, gegrillt oder gekocht. Diese Brassenart ist auch unter den namen Kakarelos, Karangiosies oder Krachantsis bekannt.

SPARAGLIONE: Pesce che preferisce le acque poco profonde e a fondo roccioso e sassoso. Comunissimo nelle acque greche.

RINGBRAXEN: En fisk som föredrar grunt vatten och klippig eller stenig botten. Den är mycket vanlig i grekiska havsvatten. Dess kött är särskilt gott på hösten. Den äts grillad, stekt eller kokt. Den kallas även Kakarellos, Karagiozis eller Krachantzis. Den mäter från 15 cm till 25 cm.

DIPLODUS ANNULARIS: Deze soort zeebrasem prefere-ert ondiep water en een rotsige of steenachtige bodem. Een in de Griekse zeeén algemeen voorkomende vissoort. Het vlees is vooral in het najaar of gekoot. Deze vis vordt in Griekenland Spáros, Kakaréllos, Karangiózis en Krachantzis genoemd. Lengte 15 tot 25 cm.

タイ科 (スパロス)

浅瀬、並びに岩や石の多い海底を好むタイ科の魚。 ギリシャ全域でよく見
られる。 肉は特に秋になるとおいしくなり、焼いたり煮たり、揚げたりして食べる。
ギリシャでは、カカレロス、カランギオジス、クラハンツィス等の名でも知られる。
体長15〜25cm.

20. *Diplodus Annularis Spáros*

SEA-BREAM FAMILY: A small-sized species of sea-bream found living in holes and fissures in rocks near to the shore, up to a depth of 25 metres. It has very tasty flesh, and can be eaten grilled or fried. Also known as a Sargi or Sargouthi. Length up to 45 cms..

ΣΑΡΓΟΣ: Είδος μικρόσωμο πού ζεῖ σέ τρύπες καί σχισμές τῶν βράχων κοντά στίς ἀκτές μέχρι βάθος 25 μ. Ἔχει πολύ νόστιμο κρέας καί γίνεται ψητός ἤ τηγανητός. Λέγεται καί Σαργί ἤ Σαργούδι. Μῆκος ὡς 15 ἑκ.

SAR COMMUN: Poisson de petite taille qui vit dans les trous et les fentes des rochers près des côtes, à une profondeur jusqu'à 25 m. En Grèce il est aussi appelé Sarghi ou Sargoudi. Longueur: jusqu'à 15 cm.

DIPLODUS SARGUS: Die 15 cm lange Abart der Meerbrasse lebt in einer Tiefe bis zu 25 m in Felslöchern und - rissen nahe der Küste. Ihr Fleisch ist sehr schmackhaft und wird gegrillt oder gebraten. Der Fisch ist auch unter den Namen Sargi oder Sargousi bekannt.

SARAGO MAGGIORE: Pesce di piccola dimensione che vive nei buchi e nelle escavazioni delle rocce vicino alle coste, ad una profondità massima di 25 metri. In Grecia viene pure chiamato "Sarhi" o "Sargoudi". Raggiunge la lunghezza di 15 centimetri.

HAVSBRAXEN-FAMILJEN: En fisk av liten storlek, upp till 15 cm lång, som lever i hål och springor i klippor nära kusterna ut till 25 m djup. Dess kött smakar utsökt, och man äter det grillat eller stekt. Den kallas även för Sarjï eller Sargoúdi.

DIPLODUS SARGUS: Een kleine soort onder de zeebrasems, die in gaten of in scheuren in rotsen bij de kust tot een diepte van 25 m. leeft. Het vlees is zeer smakkelijk en wordt gegrilld of gebakken. Deze vissoort is in Griekenland bekend als Sargós, Sargi, en Sargoédi. Lengte tot 15 cm.

タイ科 (サルゴス)

海深25m以下の沿岸の岩の割れ目や穴に住む。 タイ科の小魚であるが味は大へん良く、焼いたり揚げたりして食べる。 サルキともサルゲジとも呼ばれる。 体長15cm以下。

21. *Diplodus Sargus Sargós*

BOGUE: This species of sea-bream is plentiful in Greek waters. It forms large shoals, swimming close to rocky shores or to places where there is seaweed. It makes particularly good eating in winter, as it is then quite fleshy and also ready to spawn. It is eaten fried or grilled. Length 20 to 35 cms..

ΓΟΠΑ: Ψάρι πολύ κοινό στίς έλληνικές θάλασσες. Σχηματίζει μεγάλα κοπάδια πού μετακινοῦνται κοντά σέ βραχώδεις άκτές ή βυθούς μέ φύκια (φυκιάδες). Τό κρέας τῆς γόπας εἶναι άρκετά νόστιμο ἰδιαίτερα τό χειμώνα πού εἶναι παχιά καί άβγωμένη. Γίνεται τηγανητή ή ψητή. Μῆκος 20 ὡς 35 έκ.

BOGUE: Ce poisson se trouve en abondance dans les eaux grecques. Il vit en bancs très denses, et nage près des côtes rocheuses dans des eaux à fond d'algues. Sa chair est très savoureuse en hiver parce qu'à cette époque elle est plus grasse et, dans le cas des femelles, enrichie d'oeufs. On la consomme frite ou grillée. Longueur: 20 à 35 cm.

BARSCH: Diese Abart der Meerbrasse trifft man in großer Zahl in griechischen Gewässern. Sie tritt in großen Schwärmen in der Nähe felsiger Küste oder mit Seegras bewachsener Stellen auf. Das Fleisch dieses Fisches ist besonders schmackhaft im Winter kurz vor der Laichzeit. Er wird gebraten oder gegrillt und wird 20-35 cm lang.

BOGA: E' di casa nelle acque greche. Vive in banchi fittissimi e nuota non lontano dalle coste rocciose a fondi coperti di alghe. Nella stagione invernale la sua carne è più saporita perchè più grassa mentre le femmine si arricchiscono di uova. Viene preparato fritto e ai ferri. La sua lunghezza raggiunge i 35 centimetri.

ABORRE: En mycket vanlig fisk i grekiska vatten, som är ungefär 20 till 35 cm lång. Den bildar stora svärmar, som rör sig nära klippiga kuster eller djup med tång (tångbankar). Dess kött har ganska god smak särskilt på vintern, då den är tjock och har rom. Man äter den antingen stekt eller grillad.

BOOPS: Deze zeebrasem-soort komt in grote aantallen voor in de Griekse zeeën. Hij leeft in grote scholen dichtbij rotsige kusten of op plaatsen met veel zeewier. Het vlees is vooral in de winter buitengewoon smakelijk, wanneer deze vis dik is en de wijfjes vol kuit zijn. Hij wordt gebakken of gegrilld gegeten. Lengte 20 tot 35 cm.

タイ科　（コバ）

タイ科のこの魚はギリシャの海に豊富に見られる。 大きな群をなして岩岸や海草地帯付近を遊泳する。 冬季には身が厚くなって産卵に備えるためことさら美味になる。 揚げたり焼いたりして食べる。 体長20～35cm.

22.   *Boops Boops Gópa*

SHEEPSHEAD BREAM: This species of sea-bream frequents any type of habitat to be found in depths of between 20 and 30 metres. Its flesh is sparse but appetizing, and is usually served fried. This fish is also known by the Greeks as Ougena. Length up to 40 cms..

ΧΑΡΑΚΙΔΑ: Εἶδος πού ζεῖ σέ κάθε λογῆς βυθό σέ βάθος 20 ὥς 30 μ. Τό κρέας του ἄν καί λιγοστό εἶναι πολύ νόστιμο. Γίνεται τηγανητό. Λέγεται καί Οὔγενα. Μῆκος μέχρι 40 ἑκ.

SAR TAMBOUR: Cette espèce fréquente toute sorte d'habitats. Le peu de chair dont elle dispose est appétissante et, généralement, se consomme frite. Ce poisson est aussi appelé Oughena par les Grecs. Sa longueur atteint 40 cm.

CHARAX PUNTAZZO: Der bis zu 40 cm lang werdende Fisch ist überall in einer Tiefe von 20-30 m anzutreffen. Er hat wenig aber schmackhaftes Fleisch und wird gebraten gegessen. In Griechenland wird er auch Ugena genannt.

SARAGO PIZZUTO: Pesce che si trova un pò dappertutto. La sua carne, benché scarsa, è succulenta. Generalmente si consuma fritto. In Grecia viene pure denominato "Oughena". La sua lunghezza raggiunge i 40 centimetri.

CHARAX PUNTAZZO: Denna sorts djuphavsbraxen lever på vilket djup som helst ifrån 20 till 30 m. Fastän den har litet kött, smakar den mycket gott. Den serveras stekt. Kallas även för Oujena. Dess längd är upp till 40 cm.

CHARAX PUNTAZZO: Een zeebrasem-soort, die elke omgeving op diepten van 20 tot 30 m. voor lief neemt. Hij heeft weinig vlees, dat echter zeer goed smaakt en gebakken wordt gegeten. In Griekenland heet deze vis Harakitha en ook Oégena. Lengte tot 40 cm.

タイ科 (ハラキサ)

タイ科のこの魚は。海深20~30mのいかなる環境にても生棲する。 肉は薄いが食欲をかりたてる。 ふつうは揚げて食べる。 ギリシャでは ウガナという名でも知られる.

23. *Charax Puntazzo Harakitha*

SAUPE: This species lives in small shoals where the bottom is rocky, or in areas well-covered in seaweed. The flesh is not particularly tasty, but may be eaten fried or boiled.

ΣΑΛΠΑ: Είδος πού ζεῖ σέ ὁμάδες στούς βραχώδεις βυθούς ἤ σέ μέρη σκεπασμένα μέ φύκια (φυκιάδες). Τό κρέας του ὄχι πολύ νόστιμο. Γίνεται τηγανητό ἤ βραστό.

SAUPE: Cette espèce vit en petits bancs en eau à fond de roches ou à fond de sable bien recouvert par des algues. Sa chair pas particulièrement savoureuse, peut être consommée frite ou bouillie.

BOOPS SALBA: Diese Art lebt in kleinen Schwärmen und bevorzugt felsigen oder stark mit Seegras bewachsenen Meeresboden. Das Fleisch ist nicht sehr schmackhaft aber eßbar.

SALPA: Questa specie vive in banchi ridotti in acque a fondo roccioso o sabbioso coperto di alghe. Pesce particolarmente saporito, viene servito fritto o lesso.

BOOPS SALPA: Denna sort lever i stora flockar på klippiga djup eller på platser täckta med tång (tångbankar). Dess kött är inte så gott. Den ätes stekt eller kokt.

BOOPS SALPA: Een in kleine scholen in zeewater met een rotsbodem of met wieren bedekte bodem levende vissoort. Het vlees is niet bijzonder smakelijk, en wordt gebakken of gekookt gegeten.

タイ科 （サルパ）

岩底や海草地帯に小さな群をなして棲息する。　さほどおいしいとは言えない
が、揚げたり煮たりすれば、食用になる。

24. *Boops Salpa Sálpa*

PICAREL: A small fish which is to be found abundantly in Greek waters, living in large shoals in depths of 6 to 30 metres. Marithes are generally eaten fried, the smaller ones in particular being in great demand as delicacies. Length up to 15 cms..

ΜΑΡΙΔΑ: Μικρό ψάρι πολύ κοινό στίς έλληνικές θάλασσες. Σχηματίζει μεγάλα κοπάδια σέ βάθος 6 ὡς 30 μ. Γίνεται τηγανητό. Ἰδιαίτερα οἱ πολύ μικρές ἀποτελοῦν ἐκλεκτό μεζέ. Μῆκος μέχρι 15 ἑκ.

PICAREL: Petit poisson très commun dans les mers grecques, où il vit en de bancs très denses se tenant à une profondeur de 6 à 30 mètres. Les picarels sont servis frits, et surtout les petits sont très recherchés comme des délicatesses. Longueur: jusqu'à 15 cm.

STINT: Ein kleiner, bis zu 15 cm langer Fisch, der sehr häufig in griechischen Gewässern zu finden ist. Er lebt in großen Schwärmen in einer Tiefe von 6-30 m. Stinte werden gebraten, besonders die kleinen Fische gelten als Delikatesse.

ZERRO: Pesciolino comunissimo nelle acque greche, in cui vive in banchi fittissimi ad una profondità che varia dai 6 ai 30 metri. Viene servito fritto e i più piccoli sono particolarmente apprezzati. La sua lunghezza raggiunge i 15 centimetri.

SMÅFISK: En liten fisk som är mycket vanlig i grekiska vatten. Den bildar stora svärmar på 6 till 30 m djup. Man steker den. Särskilt de mycket små utgör en utsökt delikatess, som tillhugg. Den blir upp till 15 cm.

MAENA SMARIS: Een in de Griekse zeeën zeer algemeen vòorkomende vissoort, die in grote scholen op een diepte van 6 tot 30 m. leeft. Hij wordt als regel gebakken gegeten. Vooral de kleine exemplaren zijn zeer gezocht als een delicatesse. Lengte tot 15 cm.

シラウオ科　（マリダ）

ギリシャの海に豊富に見られる小魚で、海深6~30mの所に大きな群をなして
一般に油で揚げて食べるが、小さい物の方が美味であるために特に需要
体長15cm以下.

25. *Maena Smaris Maŕitha*

SPRAT FAMILY: A species similar to maena smaris, but bigger. Lives at a depth of 10 to 20 metres, over seaweed or close to rocks. Though eaten fried, it is not a particularly tasty fish. Length up to 25 cms..

ΜΕΝΟΥΛΑ: Ψάρι παρόμοιο μέ τή μαρίδα ἀλλά πιό μεγάλο. Ζεῖ σέ βάθος 10 ὥς 20 μ. πάνω ἀπό βυθούς μέ φύκια, κοντά σέ βράχους. Τρώγεται τηγανητό ἀλλά δέν εἶναι πολύ νόστιμο. Μῆκος ὥς 25 ἑκ.

MENDOLE COMMUNE: Cette espèce ressemble a Maena Smaris, en plus grand. Les individus vivent à une éprofondeur de 10 à 20 mètres, sur les algues ou une près des rochers. On le mange frit, mais sa chair n'est pas très savoureuse. Sa longueur atteint 25 cm.

MAENA MAENA: Eine Abart die dem Stint ähnelt, jedoch etwas größen ist. Sie lebt in einer Tiefe von 10-20 m über Seegras oder in der Nähe von Felsen. Der Fisch ist eßbar aber nicht schmackhaft.

MENNOLA: Specie che rassomiglia alla Maena Smaris, ma di dimensioni maggiori. Pesce che vive a una profondità che varia tra i 10 e i 20 metri, tra le alghe o vicino alle rocce. Viene consumato fritto, tuttavia la sua carne non è particolarmente saporita. Raggiunge la lunghezza di 25 centimetri.

VANLIG LAXERFISK: En fisk som liknar den föregående maridha mycket, men den är större. Den lever på djup mellan 10 och 20 m över tångbotten, nära klippor. Man äter den stekt, men den är inte så god. Den blir upp till 25 cm.

MAENA MAENA: Deze soort vis lijkt op de Maena Smaris (Maritha) maar is groter. Hij leeft op een diepte van 10 tot 20 m. boven zeewieren of bij rotsen. Hij wordt gebakken gegeten, maar is niet erg smakelijk. Lengte tot 25 cm.

シラウオ科 （マヌーラ）

前述の マリタに似ているがもっと大きい。 海深10〜20mの海草地帯や岩の多い所に住む。 揚げて食べるが、さしておいしい魚ではない。 体長25cm以下。

26. *Maena Maena Ménoula*

JEWFISH: A species of jewfish found in small shoals in shallow waters near the shore. Inhabits areas with a sandy or rocky bottom. Generally served fried or boiled. Also known in Greece as Kaliakoutha. Length 30 to 50 cms..

ΜΥΛΟΚΟΠΙ: Είδος πού σχηματίζει μικρές ομάδες στά ρηχά μέρη κοντά στίς ακτές, σέ βυθούς αμμουδερούς ή πετρώδεις. Τρώγεται τηγανητό ή βραστό. Λέγεται καί Καλιακούδα. Μήκος 30 ώς 50 έκ.

OMBRINE: Poisson qui forme des bancs peu nombreux en eau peu profonde près de la côte. Il préfère les eaux à fond de sable ou de pierres. Généralement on le mange frit ou bouilli. En Grèce il est également connu sous le nom de Caliacouda. Longueur: 30 à 50 cm.

UMBRINA CIRRHOSA: Dieser Fisch lebt in kleinen Schwärmen in seichten Küstengewässern und bevorzugt sandigen oder felsigen Grund. Er wird 30-50 cm lang, gebraten oder gekocht und ist in Griechenland auch unter dem Namen Kaliakoytha bekannt.

OMBRINA: Pesce che vive in banchi poco numerosi ed in acque poco profonde a fondi sabbiosi o sassosi. Generalmente viene servito lesso o fritto. In Grecia è anche noto sotto la denominazione di "Calliacouda". La sua lunghezza varia tra i 30 e i 50 centimetri.

UMBER: En fisk som bildar små flockar på grunt vatten nära kusten, med sandig botten eller stenbotten. Man äter den stekt eller kokt. Den kallas även Kaliakouda. Den blir mellan 30 och 50 cm.

UBRINA CIRRHOSA: Deze soort vis leeft in kleine scholen in ondiep water bij de kust op plaatsen met een zand- of steenbodem. Hij wordt als regel gebakken of gekookt gegeten. In Griekenland is deze vis bekend onder de namen Milokópi en Kaliakoéda. Lengte van 30 tot 50 cm.

ハタ科 (ミロコピ)

沿岸の浅瀬に小さな群をなして見られるスズキの一種　砂底や岩底のある所
に住む。　一般に揚い
呼ばれる。　体長 30 〜

58

27. *Umbrina Cirrhosa Milokópi*

CORB: An in-shore species, the Corb prefers to stay where the bottom is sandy. It usually moves around at night and hides during the day. The flesh has a pleasant flavour, and is served cooked in a variety of ways. The Corb is also known as the Sikiós or Marioli. Length up to 70 cms.

ΣΚΙΟΣ: Ψάρι τῶν ἀκτῶν πού προτιμᾶ τούς ἀμμουδερούς βυθούς. Κυκλοφορεῖ κυρίως τή νύχτα καί κρύβεται τήν ἡμέρα. Τό κρέας του ἀρκετά νόστιμο. Τρώγεται μέ πολλούς τρόπους. Λέγεται καί Σικιός ἤ Μαριόλι. Μῆκος μέχρι 70 ἑκ.

CORB OU CORBEAU: Ce poisson préfère les côtes et les eaux à fond de sable, où il reste pendant le jour: car son activité a lieu surtout pendant la nuit. Sa chair a bon goût et on la prépare de plusieurs façons. Le corbeau en Grèce est également connu comme Sikios ou Marioli. Longueur: jusqu'à 70 cm.

CORVINA NIGRA: Ein in Küstengewässern aufretender Fisch, der sandige Gründe bevorzugt. Er bewegt sich dei Nacht und versteckt sich bei tag. Sein Fleisch ist schmackhaft und er wird auf verschiedene Arten zubereitet. In Griechen-land ist dieser bis zu 70 cm lang werdende Fisch auch unter den Namen Sikio oder Marioli bekannt.

CORVO: Pesce che preferisce le coste e le acque poco profonde a fondo di sabbia ove trascorre le ore del giorno siccome la sua attività si svolge di notte. La sua carne è saporita. Viene preparato in diversi modi. E' conosciuto in Grecia anche sotto la denominazione di "Sikios" o di "Marioli". Raggiunge lunghezza di 70 centimetri.

CORVINA NIGRA: En kustvattensfisk som föredrar sandig botten. Den simmar omkring framför allt på natten och gömmer sig på dagen. Dess kött smakar tämligen gott. Den äts på många sätt tillagad. Kallas även för Sikios eller Marioli. Dess längd är upp till 70 cm.

CORVINA NIGRA: Een bewoner van kustwater met een zanbodem. Hij is vooral 's nachts actief. Overdag houdt hij zich schuil. Het vlees is vrij smakelijk en wordt op verschillende manieren toebereid. Hij wordt in Griekenland Skiós, Sikiós en Marióli genoemd. Lengte tot 70 cm.

スキオス（フキオス）

砂底を好む近海魚。ふつうは夜行性で昼間はかくれている。肉はうまく、いろいろな料理法で食べられる。シキオスともマリオリとも呼ばれる。体長70 cm以下。

28.   *Corvina Nigra Skiós*

WRASSE FAMILY: A very common member of the Wrasse family which can be found in moderate depths of up to 100 metres. There are many similar species bearing the same name. The colours vary from area to area, and the species are therefore extremely difficult to differentiate. Length between 15 and 50 cms.. Not a particularly tasty fish to eat.

XEIΛOΥ: Πολύ κοινό ψάρι πού ζεῖ σέ μέτριο βάθος, ὡς 100 μ. Ὑπάρχουν πολλά παρόμοια εἴδη μέ τό ἴδιο ὄνομα πού τό χρῶμα τους ποικίλλει ἀπό τόπο σέ τόπο· γι᾿ αὐτό δύσκολα τό ξεχωρίζουμε. Τό μῆκος τους κυμαίνεται ἀπό 15 ὡς 50 ἑκ. Τό κρέας τους μέτριο.

MERLE: Poisson très commun qui habite des profondeurs allant jusqu'à 100 mètres. Il y a beaucoup d'espèces voisines de celle-ci avec le même nom, et dont la couleur varie suivant les lieux: c'est pourquoi on les distingue difficilement les uns des autres. Leur longueur varie entre 15 et 50 cm. Leur chair est médiocre.

LABRUS MERULA: Ein sehr verbreitetes, 15-50 cm langes Mitglied der Brassenfamilie, das in Tiefen bis zu 100 m zu finden ist. Es gibt viele ähnliche Arten des gleichen Namens. Die Farben differieren gebietsweise, sodaß es sehr schwer ist, die einzelnen Arten voneinander zu unterscheiden. Der Fisch ist eßbar aber nicht schmackhaft.

TORDO NERO: Pesce comunissimo che vive in profondità che possono raggiungere i 100 metri. Vi sono numerose specie affini dello stesso nome il cui colore varia a seconda dei luoghi, ragion per cui non è facile riuscire a distinguerle. La loro lunghezza oscilla tra i 15 ed i 50 centimetri. La sua carne non sa di molto.

LABRUS MERULA: En mycket vanlig fisk som lever på medeldjup ned till 100 m. Det finns många liknande slag med samma namn, vars färg dock skiftar från plats till plats och därför svår att särskilja. Dess längd växlar mellan 15 och 50 cm. Deras kött är medelmåttigt.

ECHTE LIPVIS: Een zeer algemeen voorkomende soort brasem, tot 100 m. diepte. Er bestaan veel aanverwante soorten onder dezelfde naam, waarvan de kleuren van plaats tot plaats verschillen, zodat het moeilijk is de diverse variëteiten van elkaar te onderscheiden. Het vlees is matig van smaak. Lengte 15 tot 50 cm.

ヘラ科 (ヒル-)

海深100m以下の程良い深さのところに見られるきわめて一般的なヘラ科の魚。同じ名前のたくさんの似通った魚がいる。 場所場所によって色が違うために判別が非常にむずかしい。 体長15~50cm。 食用としてはあまりうまいと言えない。

29. *Labrus Merula Hiloú*

RAINBOW WRASSE: A small, darting fish, the Rainbow Wrasse chooses to live in shallow waters in among seaweed and rocks. At night it rests on the bottom lying on one side. Its flesh tastes very good and it is generally eaten fried. Length up to 25 cms..

ΓΥΛΟΣ: Ψάρι μικρό πολύ εὐκίνητο πού ζεῖ στά ρηχά νερά ἀνάμεσα σέ φύκια καί πέτρες. Τό βράδυ κάθεται στό βυθό ξαπλώνοντας στό ἕνα πλευρό. Ἔχει νόστιμο κρέας καί τρώγεται τηγανητός. Μῆκος μέχρι 25 ἑκ.

GIRELLE COMMUNE: Petit poisson très agile, fréquentant les endroits peu profonds, entre les rochers et les algues. Pendant la nuit, la girelle repose au fond, couchée sur le côté. Sa chair est savoureuse et se consomme frite. Longueur: jusqu'à 25 cm.

REGENBOGENBRASSE: Ein kleiner, herumschießender Fisch, der seichte Gewässer bevorzugt und unter Seegras und Felsen lebt. Nachts ruht er auf dem Meeresboden, auf der Seite liegend. Das Fleisch dieses bis zu 25 cm lang werdenden Fisches ist sehr schmackhaft und wird gebraten gegessen.

DONZELLA: Pesciolino agilissimo. Lo si trova in acque poco profonde tra le rocce e le alghe. Durante la notte riposa sul fondo marino. Fritto è saporitissimo. La sua lunghezza raggiunge i 25 centimetri.

REGNBÅGES LÄPPFISK: En liten, mycket rörlig fisk som lever i grunt vatten mellan tång och stenar. På kvällen lägger den sig kvar på botten på ena sidan. Den har gott kött och man äter den stekt. Den mäter upp till 25 cm.

GIRELLE:Een kleine, zeer bewegelijke vis die de voorkeur geeft aan ondiep water en die tussen zeewieren en rotsen leeft. 's Nachts rust hij op zijn zijde op de bodem. Het vlees is smakelijk en wordt gebakken gegeten. Lengte tot 25 cm.

キュウセン （ギロス）.

動きのすばやい小魚で、海藻や岩の間の浅瀬を選んで住む。 夜は横にな
って海底で休む。 肉は大へんおいしく、一般に揚げて食べる。 体長25cm
以下.

LABRIDAE

30.   Coris Julis Gilos

WRASSE FAMILY: This small member of the Wrasse family is very common around Greece. It prefers shallow waters and during the day moves about continuously. At night, however, it sleeps lying on the bottom. Usually served fried. Length up to 15 cms..

ΓΥΛΑΡΙ: Εἶδος μικρόσωμο, ἄφθονο στίς ἑλληνικές θάλασσες. Τό βράδυ πλαγιάζει καί κοιμᾶται στό βυθό. Τρώγεται τηγανητό. Μῆκος μέχρι 15 ἑκ.

GOUJON: Ce poisson de petite taille est très commun dans les eaux grecques. Il préfère les eaux peu profondes et, pendant la journée, il poursuit une activité inlassable. Pendant la nuit, au contraire, il repose sur le fond. Généralement on le mange frit. Sa longueur peut atteindre 15 cm.

THALOSSOMA PAVO: Kleine Abart der Brassen, die sehr häufig in Griechenland zu frinden ist. Dieser Fisch bevorzugt seichte Gewässer und befindet sich tagsüber in ständiger Bewegung. Nachts schläft er, auf dem Meeresgrund liegend. Er wird bis zu 15 cm lang und gebraten gegessen.

GHIOZZO: Questo pesce di piccole dimensioni è comunissimo nelle acque greche. Preferisce le piccole profondità e di giorno non sta mai fermo. Di notte riposa sul fondo marino. Generalmente viene preparato fritto. La sua lunghezza raggiunge i 15 centimetri.

LÄPPFISK-FAMILJEN: Denna mycket småväxta sort finns i överflöd i grekiska vatten. På kvällen lägger även den sig att sova på botten. Den äts stekt. Dess längd blir upp till 15 cm.

PAUWLIPVIS: Een kleine vissoort die veelvuldig voorkomt in de zeeën rondom Griekenland. Hij geeft de voorkeur aan ondiep water. Overdag is hij voortdurend actief. 's nachts slaapt hij op de bodem. Hij wordt gebakken gegeten. Lengte tot 15 cm.

ベラ科 (キラリ)

ベラ科のこの小さな魚は、ギリシャ全域で大へんよく見られる。 浅瀬を好み、昼間は絶え間なく動きまわるが、夜になると海底にふして眠る。 ふつうは揚げて食べる。 体長15cm以下。

31. Thalossoma Pavo Gilári

PARROTFISH FAMILY: A species of Parrotfish, found living in shoals where there are rocky bottoms, in depths of up to 100 metres. It makes very good eating, and can be prepared in a number of different ways.

ΣΚΑΡΟΣ: Ψάρι πού ζεῖ κοπαδιαστά σέ βραχώδεις βυθούς μέχρι βάθος 100 μ. Ἔχει νόστιμο κρέας καί τρώγεται μέ πολλούς τρόπους.

POISSON-PERROQUET CRETOIS: Une variété de Poisson-Perroquet, qui vit en bancs là où il y a des fonds rocheux, à des profondeurs atteignant 100 mètres. Sa chair est très bonne, et peut être préparée de plusieurs façons différentes.

SEEPAPAGEI: Eine Abart des Papageienfisches, der in kleinen Schwärmen auf felsigem Meeresgrund lebt, in Tiefen bis zu 100 m. Sein Fleisch ist sehr schmackhaft und wird auf verschiedene Arten zubereitet.

SCARUS CRETENSIS: Vive in banchi in luoghi in cui abbondano i fondi rocciosi in profondità che raggiungono i 100 metri. La sua carne è succulenta e viene cucinata in diversi modi.

PAPEGOJ-FISK: En slags av Papegoj-fiskfamiljen, som lever i flockar på klippbotten upp till 100 m djup. Dess kött smakar mycket gott och man kan laga till den på många olika sätt.

SCARUS CRETENSIS: Deze soort papagaaivis leeft in scholen op plaatsen met een rotsbodem tot een diepte van 100 m. Zijn zeer smakelijke vlees wordt op diverse maieren toebereid.

ベラ科　(スカロス)

ベラ科のひとつで、海深100m以下の岩底地帯に群をなして主棲する。　食用にはおあつらえ向きの魚で、様々の方法で料理される。

SCARIDAE

32.   *Scarus Cretensis Skáros*

COMMON MACKEREL: The common Mackerel is found in abundance in Greek waters. In summer, it swims in large shoals near the surface. In winter it stays on the bottom, lying there motionless almost without eating. It is usually served either grilled or fried. In autumn, the Greeks salt them, whereas in spring they are dried in the sun and used to prepare "tsiro". Length up to 50 cms..

ΣΚΟΥΜΠΡΙ: Εἶδος πολύ κοινό στίς ἑλληνικές θάλασσες. Τό καλοκαίρι ζεῖ σέ μεγάλα κοπάδια κοντά στήν ἐπιφάνεια. Τό χειμώνα κατεβαίνει στό βυθό καί μένει ἀκίνητο σχεδόν χωρίς νά τρώει. Γίνεται ψητό ἤ τηγανητό. Το φθινόπωρο πού εἶναι παχύ τό παστώνουν ἐνῶ τήν ἄνοιξη τό ξεραίνουν στόν ἥλιο καί φτιάχνουν τόν "τσίρο". Μῆκος μέχρι 50 ἑκ.

MAQUEREAU COMMUN: Le maquereau se rencontre en abondance dans les eaux grecques. Pendant l'été, il nage en bancs très denses près de la surface. En hiver, il reste sur le fond, immobile, presque sans manger. Généralement, on le prépare frit ou grillé. En automne, les Grecs les salent, tandis qu'au printemps ils les sèchent au soleil pour préparer le "tsiro". Longueur: jusqu'à 50 cm.

MAKRELE: Die gewöhnliche, bis zu 50 cm lange Makrele ist in griechischen Gewässern im Überfluß zu finden. Im Sommer schwimmt sie in großen Schwärmen nahe der Meeresoberfläche. Im Winter hält sie sich auf dem Meeresgrund auf, wo sie - fast ohne Nahrungsaufnahme - bewegungslos liegt. Im Herbst, werden sie eingesalzen, im Sommer an der Sonne getrocket. Auch gegrillt oder gebraten sind sie sehr schmackhaft.

SCOMBRO COMUNE: Pesce che si trova in abbondanza nelle acque greche. Durante l'estate nuota in banchi fittissimi sfiorando la superficie marina. D'inverno rimane sul fondo, immobile, nuotando raramente. Generalmente viene preparato fritto o ai ferri. Nella stagione autunnale, quando esso è più grasso e saporito, viene salato e in primavera i greci lo mettono a seccare al sole, ottenendo cos' lo "Tsiro".

VANLIG MAKRILL: Den vanliga makrillen finner man i stora mängder i grekiska vatten. På sommaren lever den i stora svärmar nära vattenytan. På vintern simmar den ner till botten och förblir orörlig, nästan utan att äta. Man lagar till den genom att grilla eller steka den. På hösten då den är fet, saltar man in den, medan man på våren torkar den i solen och gör "tsiro".

MAKREEL: De gewone makreel komt in de Griekse wateren in grote aantallen voor. In de zomer zwemt hij in grote scholen dicht onder de oppervlakte. De winter brengt hij door op de bodem, bewegingloos en vrijwel zonder te eten. Hij wordt in het algemeen geroosterd of gebakken gegeten. In het najaar, wanneer de makrelen vleziger zijn, worden zij in Griekenland gezouten, terwijl zij in de lente in de zon worden gedroogd. Men noemt de gedroogde makrelen "tsiro". Lengte tot 50 cm.

サバ （スクブリ）

ギリシャの海に豊富に見られる。　夏は海面近く大きな群をなして遊泳するが、冬になるとほとんど物も食べずじっと海底にふしている。　ふつうは焼くか揚げるかして食べる。　最も肉のつく秋には、ギリシャ人はこれを塩漬けにし、春が来たところで太陽で乾かし、ツィロという料理に使う。　体長50cm以下。

70

33. *Scomber Scombrus Scoubrí*

CHUB MACKEREL: The Chub Mackerel move together in large shoals, swimming near the surface. In autumn, when the fish are most fleshy, the Greeks salt them, whereas in spring they are dried in the sun and used to prepare "tsiro", exactly as with the Common Mackerel. Also eaten fried, grilled or baked.

ΚΟΛΙΟΣ: Ψάρι πού σχηματίζει μεγάλα κοπάδια καί ζεῖ κοντά στήν ἐπιφάνεια τῆς θάλασσας. Τό φθινόπωρο πού εἶναι παχύς τόν παστώνουν, ἐνῶ τήν ἄνοιξη τόν ξεραίνουν καί φτιάχνουν τόν ''τσίρο'' ὅπως καί μέ τό σκουμπρί. Τρώγεται ἐπίσης τηγανητός, ψητός ἤ στόν φοῦρνο.

MAQUEREAU ESPAGNOL: Ces poissons se déplacent en bancs très denses qui nagent près de la surface. En automne, quand ces poissons sont bien gras et savoureux, les Grecs les salent, tandis qu'au printemps ils les sèchent au soleil pour préparer le "tsiro", exactement comme pour le maquereau commun. Il peut être aussi mangé frit, grillé ou cuit.

SCOMBER COLIAS: Diese Abart der Makrele schwimmt in großen Schwärmen nahe der Meeresoberfläche. Im Herbst werden die Fische eingesalzen, im Sommer an der Luft getrocknet, aber auch gegrillt oder gebraten.

LANZADO: Questi pesci si spostano in banchi fittissimi che nuotano sfiorando la superficie marina. In autunno viene salato ed in primavera messo a seccare al sole per la preparazione del noto "Tsiro", esattamente come viene fatto per lo scombro comune. Lo si mangia fritto o preparato in altri vari modi.

SPANSK MAKRILL: En fisksort som bildar stora flockar och lever nära havsytan. På hösten då den är fet, saltar man in den, och på våren torkar man den och lagar till "tsiro", som med den vanliga makrillen. Den äts även stekt, grillad eller tillagad i ugnen.

SPAANSE MAKREEL: Deze makreelsoort zwemt in grote scholen dicht onder de oppervlakte. In het najaar, wanneer zij veel vlees hebben, worden ze in Griekenland gezouten, terwijl zij in de lente worden gedoogd. Zij worden gedroogd "tsiro" genoemd, evenals de gedroogde gewone makrelen. Hij wordt ook gebakken en geroosterd, of in de oven bereid.

サバ （コリオス）

海面近く大きな群れをなして遊泳する。　秋になって身が厚くなると、前述のスクフリと全く同じように、ギリシャ人はこれを塩漬けにして、春が来た時に太陽で乾燥させて ツィロという料理に使う。　揚げたり焼いたりする他、オーブン料理にも使われる。

34. Scomber Colias Koliós

TUNA: One of the largest of the migrating fish, the Tuna frequents northern waters during the summer months. In winter, however, it travels down into the Mediterranean, and remains there living at a great depth. In spring, it comes in to the shore, spawns, and departs in shoals for the north. The flesh is both tasty and plentiful, but is only to be found tinned. Length up to 4 metres.

ΤΟΝΟΣ: Πολύ μεγάλο μεταναστευτικό ψάρι. Τό καλοκαίρι ζεῖ στίς βόρειες θάλασσες ἐνῶ τό χειμῶνα κατεβαίνει στή Μεσόγειο καί ζεῖ σέ μεγάλο βάθος. Τήν ἄνοιξη πλησιάζει στίς ἀκτές, γεννᾶ καί φεύγει κοπαδιαστά γιά τό Βοριά. Τό κρέας του νόστιμο καί παχύ γίνεται προπάντων κονσέρβες. Μῆκος μέχρι 4 μ.

THON ROUGE: Un des plus grands poissons migrateurs, le thon pendant les mois d'été fréquente les eaux nordiques. Mais en hiver, il descend en Méditerranée où il vit aux grandes profondeurs. Au printemps, il s'approche des côtes pour la ponte, puis il repart, en bancs immenses, pour le Nord. Sa chair, grasse et succulente, se trouve surtout en conserves. Longueur: jusqu'à 4 mètres.

THUNFISCH: Einer der größten wandernden Fische. Er hält sich während der Sommermonate in nördlichen Gewässern auf. Im Winter wandert er ins Mittelmeer und lebt dort in großen Tiefen. Im Frühling kommt er an die Küste, laicht dort und zieht in großen Schwärmen nach Norden zurück.

TONNO: Uno dei maggiori pesci migratori. Pesce che durante i mesi estivi frequenta le acque nordiche e che nella stagione invernale scende nel Mediterraneo ove vive in grandi profondità. All'apparire della primavera si avvicina alle coste per deporvi le uova, poi ne riparte, in banchi immensi, per il Nord. La sua carne grassa e succulenta viene venduta al commercio in conserva. La sua lunghezza raggiunge i 4 metri.

TONFISK: En mycket stor flyttfisk. På sommaren lever den i haven i norr, medan den på vintern flyttar ner till Medelhavet och där på stora djup. På våren närmar den sig kusterna, föder och ger sig iväg i stora flockar mot norr. Dess kött är mycket gott och fett och man konserverar det först och främst. Dess längd blir upp till 4 m.

TONIJN: De tonijn is een van de grootste soorten trek vissen. In de zomer leeft hij in de Noordelijke zeeën, maar in de winter trekt hij naar de Middellandse Zee waar hij op grote diepte leeft. In de lente komt hij dicht bij de kust, schiet kuit, en trekt dan in grote scholen terug naar het Noorden. Zijn smakelijk en overvloedig vlees wordt hoofdzakelijk tot conserven verwerkt. Lengte tot 4 m.

マグロ (トンノス)

最も大型の回遊魚のひとつ。 夏の間は北の海に住むが 冬になると1也中海へ 下ってきて深いところに生棲する。 春が来ると沿岸に近づき産卵し 群をなしてまた北にする。 肉は美味であると共に量も多いが カン詰としてしか出まわらない。 は長4m以下。

74

35. *Thunnus Thynnus Tónnos*

STRIPED GREY MULLET: A shallow-water species found near the seashore. Usually frequents harbours, salt-water lakes, river-mouths, etc.. Adapts easily to fresh-water living, and often enters stagnant and infected waters. The flesh is very tasty, provided that it is caught in unpolluted waters. Usually eaten boiled or fried.

ΚΕΦΑΛΟΣ: Ψάρι τῶν ρηχῶν νερῶν τῆς ἀκτῆς. Τό βρίσκουμε συχνά μέσα στά λιμάνια, στίς λιμνοθάλασσες, στίς ἐκβολές ποταμῶν κ.λ.π. Προσαρμόζεται καί ζεῖ ἄνετα καί στό γλυκό νερό. Συχνά πηγαίνει σέ νερά στάσιμα καί μολυσμένα. Τό κρέας του πολύ νόστιμο, ἀρκεῖ νά προέρχεται ἀπό καθαρό μέρος. Τρώγεται βραστός ἤ τηγανητός.

MULET CABOT: Espèce d'eau peu profonde, que l'on rencontre près des côtes. Généralement, le mulet cabot fréquente les ports, les lacs salés, les embouchures des fleuves, etc. Il s'adapte facilement à l'eau douce, et souvent il s'aventure dans d'eaux stagnantes ou infectes. Sa chair est très bonne, pourvu qu'il soit pêché dans des eaux non-polluées. Il est généralement frit ou bouilli.

MEERASCHE: Ein Fisch, der in seichtem Wasser nahe der Küste lebt. Gewöhnlich ist er in Häfen, Salzwasserseen, Flußmündungen usw. zu finden. Er paßt sich mit Leichtigkeit den Lebensbedinungen im Süßwasser an und schwimmt oft in stehenden oder verschmutzten Gewässer. Sein Fleisch ist sehr schmackhaft, vorausgesetzt, daß der Fisch in klarem Wasser gefangen wurde. Er wird gekocht oder gebraten gegessen.

MUGGINE-CEFALO: Pesce che vive in acque poco profonde, vicino alle coste. Generalmente lo si trova nelle acque dei porti, dei laghi salati, ecc. Si adatta facilmente all'acqua dolce e non di rado si avventura fino a raggiungere acque stagnanti e infette. La sua carne è ottima. Viene preparato lesso o fritto.

RANDIG MULTEFISK: En fisk som finns vid kusternas grunda vatten. Man fiskar den ofta inne i hamnarna, i havssjöar och vid floders utflöde etc. Den vänjer sig och lever även utan svårighet i sötvatten. Den simmar ofta omkring i vatten, som står still och är förorenat. Dess kött är mycket gott, om man bara fiskat den på en ren plats. Man äter den kokt eller stekt.

HARDER: Een vis van het ondiepe water bij de kust. Gewoonlijk komt hij voor in havens, zoute meren, riviermondingen enz. Hij past zich makkelijk aan in zoet water. Vaak leeft hij ook in stilstaand of vervuild water. Zijn vlees is zeer smakelijk, mits hij niet in vervuild water is gevangen. Hij wordt gekookt of gebakken gegeten.

マボラ （ケファロス）

海岸近くに見られる浅瀬の魚。 ふつう港や塩水湖 河口等にしばしは見ら
れる。 淡水にも簡単に適応し しばしは汚水の中にも入りこむ。 水のきれい
な所でつかまえさえすれは大へんにおいしい。 ふつうは煮たり揚げたりして食べる

36. *Mugil Cephalus Kéfalos*

SCORPION FISH FAMILY: A species of scorpion-fish which chooses to live near rocky bottoms in depths of 20 to 50 metres. It has delicious white flesh, which is served in soup. Length up to 50 cms..

ΣΚΟΡΠΙΝΑ: Είδος πού ζεῖ στούς βραχώδεις ὂυθούς σέ βάθος 20 ὥς 50 μ. ῎Εχει πολύ νόστιμο καί ἄσπρο κρέας. Τρώγεται ὂραστή σούπα. Μῆκος ὥς 50 ἐκ.

RASCASSE ROUGE: Il s'agit d'une variété méditerranéenne du Poisson-Scorpion des Antilles, qui préfère vivre près des fonds rocheux, se tenant à des profondeurs de 20 à 50 mètres. Sa chair, blanche et succulente, fait une très bonne soupe. Longueur: jusqu'à 50 cm.

SKORPIONFISCH: Diese Abart des Skorpionfisches bevorzugt felsige Meeresböden mit einer Tiefe von 20-50 m. Der bis zu 50 cm lang werdende Fisch hat sehr schmackhaftes weisses Fleisch, aus dem Suppe gekocht wird.

SCORFANO ROSSO: Trattasi d'una varietà mediterranea del pesce delle Antille, che preferisce vivere nelle vicinanze dei fondi rocciosi in profondità dai 20 fino ai 50 metri. La sua carne è bianca e succulenta: se ne fa un'ottima zuppa. Raggiunge la lunghezza di 50 centimetri.

SKORPION-FISKFAMILJEN: En slags Skorpion-fisk som lever på klippig botten på djup mellan 20 och 50 m. Den har ett mycket välsmakat och vitt kött. Man äter den i fisksoppa. Den blir upp till 50 cm.

SCHORPIOENVIS: Deze soort schorpioenvis heeft voorkeur voor een rotsachtige bodem op een diepte van 20 tot 50 m. Hij heeft zeer goed smakend wit vlees, waarvan soep wordt bereid. Lengte tot 50 cm.

フサカサゴ科 (スコルピナ)

海深20〜50mの岩底地帯近くを選んで住むカサゴ族のひとつ。 肉は白身でおいしく スープに使われる。 体長50cm以下.

37. *Scorpeana Scrofa Skorpina*

SCORPION FISH FAMILY: Similar to scorpaena scrofa but much smaller, this species is greyish-brown in colour, and inhabits rocky areas. It has delicious flesh, and is eaten boiled. Length up to 30 cms..

ΣΚΟΡΠΙΟΣ: Παρόμοιος μέ τήν Σκορπίνα άλλά πολύ μικρότερος καί μέ χρῶμα σταχτί-καφέ. Ζεῖ σέ βραχώδεις βυθούς. ῎Εχει πολύ νόστιμο κρέας καί τρώγεται βραστός. Μῆκος ὡς 30 ἑκ.

RASCASSE NOIRE: Ce poisson ressemble, en plus petit, à Scorpaena Scrofa; sa couleur est entre le gris et le brun, et il habite les emplacements rocheux. Sa chair bouillie est délicieuse. Longueur: jusqu'à 30 cm.

SCORPAENA PORCUS: Diese Art ist der vorhergegangenen sehr ähnlich, aber viel kleiner. Sie lebt in felsigen Gründen und ihre Farbe ist graubraun. Das Fleisch des bis zu 30 cm lang werdenden Fisches ist sehr schmackhaft und wird gekocht gegessen.

SCORFANO NERO: Questo pesce, benché più piccolo, somiglia alla Scorpaena Scrofa; Il suo colore oscilla tra il grigio e il bruno. Vive in luoghi rocciosi. Servito lesso è succulento. La sua lunghezza raggiunge i 30 centimetri.

SKORPION-FISKFAMILJEN: Den liknar mycket den föregående Skorpios-fisken, men är mycket mindre och har en gråbrun färg. Den lever vid klippiga djup. Den har ett mycket gott kött och man äter den kokt. Längd upp till 30 cm.

SCHORPIOENVIS: Deze schorpioenvis lijkt op de Scorpaena Scrofa, maar is veel kleiner en is grijs-bruin van kleur. Hij leeft op rotsige bodems. Zijn uitstekend smakend vlees wordt gekookt gegeten. Lengte tot 30 cm.

フサカサゴ科 （スコルピ°オス）

前述のスコルピ°ナに似ているがもっと小さい。　色は灰色がかった茶色で岩の多い
所に住む。　肉は美味で煮て食べる。　体長30cm以下。

SCORPAENIDAE

38. *Scorpaena Porcus Skorpiós*

STREAKED GURNARD: A species of gurnard which lives near sandy or muddy bottoms, close to the shore. Very common, in the Aegean Sea. The flesh is white and very tasty, and is served boiled as a soup.

ΚΑΠΟΝΙ: Ψάρι πού ζεῖ στούς ἀμμώδεις ἤ λασπώδεις βυθούς κοντά στίς ἀκτές. Εἶδος πολύ κοινό στό Αἰγαῖο. Τό κρέας του ἄσπρο καί πολύ νόστιμο. Τρώγεται βραστό σούπα.

GRONDIN IMBRIAGO: Ce poisson vit près des fonds boueux ou sablonneux, près de la côte. Très commun dans la mer Egée. Sa chair est blanche et très bonne: on la sert bouillie, dans sa propre soupe.

TRIGLA LINEATA: Es handelt sich um eine Abart der Triglidenfische, die in der Nähe sandiger oder schlammiger Gründe nahe der Küste leben. Im Agäischen Meer sind sie sehr häufig zu finden. Das Fleisch ist weiss und sehr schmackhaft und wird in Suppe serviert.

CAPPONE DALMATO: Pesce che vive in acque melmose e sabbiose, vicino alle coste. Comunissimo nel Mar Egeo. La sua carne è bianca e bottima. Viene servito lesso in zuppa.

RANDIG KNORRHANE FISK: En fisk som lever vid sandig eller lerig botten nära kusten. Den är mycket vanlig i Egeiska havet. Dess vita kött smakar mycket gott. Man äter den kokt i soppa.

POON: Een vissoort die op zanderige o:f modderige bodems leeft dicht bij de kust. Een zeer gewone vis in de Aegeᵗsche Zee. Het vlees is wit en zeer smakelijk, en wordt in soep gegeten.

ホウボウ科 （カポニ）

毎岸に近い砂底や泥底に住むホウボウの一種。 エーゲ海では大へん一般 的である。 肉は白身で非常にうまく、煮てスープに使われる。

39. *Trigla Lineata Kapóni*

SOLE: The Sole frequents flat, sandy or pebbly bottoms, in depths of more than 80 metres. Changes its colour naturally to always resemble its surroundings, and to camouflage itself. He flesh is lean and tasty, and is usually served grilled. Length 50 cms at most.

ΓΛΩΣΣΑ: Ψάρι πού ζεῖ σέ ὁμαλούς βυθούς μέ ἄμμο ἤ χαλίκια σέ βάθος μεγαλύτερο ἀπό 80 μ. Ἀλλάζει εὔκολα τό χρῶμα του γιά νά γίνεται κάθε φορά ὅμοιο μέ τό περιβάλλον καί νά καμουφλάρεται. Ἔχει κρέας νόστιμο καί ἄπαχο. Γίνεται συνήθως ψητό ἤ τηγανητό. Μῆκος μέχρι τό πολύ 50 ἑκ.

SOLE: La sole fréquente les fonds de sable ou de cailloux, se tenant à des profondeurs qui dépassent 80 mètres. Elle change facilement de couleur pour s'adapter à son entourage naturel et pour se camoufler contre ses ennemis. Sa chair, maigre et savoureuse, est généralement servie frite ou grillée. Longueur: 50 cm au maximum.

SEEZUNGE: Die Seezunge lebt auf sandigen oder kiesigen Meeresböden in einer Tiefe von 80 m. Farbveränderungen entsprechen immer der Umgebung und dienen der Tarnung. Das Fleisch ist mager und schmackhaft und wird gegrillt oder gebraten gegessen. Die Seezunge wird bis zu 50 cm lang.

SOGLIOLA: Pesce che frequenta i fondi sabbiosi e sassosi, mantenendosi a profondità non superiori agli 80 metri. Cambia facilmente colore per adattarsi all'ambiente e più facilmente ingannare i suoi nemici. La sua carne, benché magra, è saporita e generalmente viene servita fritta o ai ferri. Raggiunge la lunghezza di 50 centimetri.

TUNGA: En fisk, som man finner på slät botten med sand eller småsten på djup över 80 m. Den skiftar lätt färgen för att varje gång likna omgivningen och på så sätt gömma sig. Den har ett gott och inte fett kött. Man lagar vanligtvis till den grillad eller stekt. Den blir inte över 50 cm lång.

TONG: De tong leeft op vlakke zand- of kiezelbodems op diepten van meer dan 80 m. Hij wisselt gemakkelijk van kleur, en neemt altijd de kleur van zijn omgeving aan om zich te camoufleren. Het vlees is niet vet, en is zeer smakelijk, en wordt als regel geroosterd of gebakken. Maximale lengte 50 cm.

シタヒラメ （グロッサ）

海深80m以上の砂や小石の平らな海底に住む。 まわりの環境に似せてから だの色を変え身を守る。 肉はうすいが美味で、ふつう焼いたり揚げたりして食べ る。 体長は大きいもので50cm.

40.   *Solea Solea Glóssa*